Barnstaple's Vanished Lace Industry

Peter Christie

and

Deborah Gahan

EDWARD GASKELL *publishers*
DEVON

First published 1997 by
EDWARD GASKELL Publishers
6 Grenville Street
Bideford • Devon
EX39 2EA

Reprinted January 2000

ISBN 1 -898546 - 21 - 5

 Recycled paper

BARNSTAPLE'S VANISHED LACE INDUSTRY

by

Peter Christie and Deborah Gahan

Typeset, printed & bound by
The Lazarus Press
Unit 7 Caddsdown Business Park
Bideford
Devon EX39 2HN

Contents

Illustrations

Between pages 64 & 65

Foreword

The story of how I became involved in writing this book begins with my childhood in the 1960s. Every Sunday I would visit my grandmother in Gaydon Street and walk through the recently cleared area where the Derby houses had stood. The old street patterns were still visible and my father would ask me to choose which one I wanted to walk down that day. In 1969 I went for my first school trip, an unusual event at that time, to Derby Lace Factory, then owned by Small and Tidmas. I remember the blackness of the old cotton machinery, the heat of the nylon room and, most of all, two old ladies sitting by themselves in a room away from the main part of the factory, mending nets. To a child of ten they seemed to have been there forever, like characters from a fairy tale.

Many years later I had to choose a project for an Open University course. In the intervening years I had heard much of the myth about Derby, which despite its reputation still holds a fond place in the hearts of many Barnstaple people. I decided to try and find out more about the factory and the area around it, what was true and whether this reputation was deserved. I had no thought of publication, although I had always wanted to write a book of some sort. My course tutor, Peter Christie, then suggested that we write an academic paper together on the lace industry in Barnstaple. This grew into the book which you now hold and for which I remain indebted to him in enabling one of my dreams to come true.

Deborah Gahan

§

v

Foreword

In 1983 I set up a Manpower Services Commission sponsored scheme whereby some fourteen previously unemployed people spent varying periods of time over two years indexing the North Devon Journal from 1850 onwards. As I checked and cross-referenced their work I noticed mentions of the lace trade in Barnstaple. When I searched the vast resources of the North Devon Athenaeum to find out more about these factories I was surprised to find so little written about what had been the most important industry in the town in the nineteenth century. Over the next 10 years I read and indexed the Journal from its inception in 1824 until 1850 looking especially for lace references. The results of that work are before you now. The history of businesses is rarely written yet they are a vital ingredient of all our ancestors' lives - we trust you will enjoy reading this contribution to that history as much as we enjoyed researching and writing it.

Peter Christie

§

Acknowledgements

We would like to acknowledge the help of the staff of both the North Devon Athenaeum, Les Franklin and Marjorie Snetzler, and the North Devon Record Office under archivist Tim Wormleighton in accessing the records used in this study. We also wish to thank David Dalziel for his account of the present state of the Derby factory and S&T for their generous support.

Throughout this book material presented in italic type represents a direct quotation from the original sources. Each quotation is numbered and references will be found at the end of the text.

Preface

If you take a walk today, out of the town centre of Barnstaple, and along Vicarage Street to where it opens out and becomes a car park and the urban relief road, it is hard to imagine that here was once a self-contained community of hundreds of houses with its own shops and pubs. Indeed for the first part of its history Vicarage Street, originally Vicarage Lane, was the only link between this area and the town.

The Victorian terraced houses which stand halfway along on the right of this street were built in the latter part of the last century on the grounds of the former Vicarage Lawn House, and so are relatively new additions. Next along from these is an older terrace which finishes with what is now a Chinese takeaway but was for many years Derby fish and chip shop. These buildings, together with the remaining corner of Higher Maudlin Street on the opposite side of Vicarage Street, mark the beginning of the Derby area. The shop forms the corner of one side of New Buildings, a cul-de-sac which leads to the old gates of the now demolished St Mary Magdalene Church. A complex of old people's housing now occupies the site of the church but the old graveyard has been retained as a grassed area. Some of the gravestones stand against the walls, and you can still walk through from the gates at New Buildings to the gateway into Bear Street by the old church path.

On the other side of the road are the surviving houses of Higher Maudlin Street with its link to what was Lower Maudlin Street. These old streets are separated from their former neighbours of Portland Buildings and the remaining part of Princes Street, now known as Princess Street, by the relief road which dissects the area, its route following that of the further side of Gaydon Street, cutting through the site of the old Derby streets and of the ancient Rackfield.

Just past New Buildings would have been the entrance to Boden's Row later known as Corser Street. The remaining Derby streets ran parallel to this but were linked at their far ends. These were, in order going away from the town, Reform Street, Union Street and Newington Street.

The Lace Factory itself, along with the Union Inn and the former Miller Institute, still stands. The main factory building has stayed remarkably unchanged to date, however, its once four storeyed splendour has been sadly diminished by the demolition of most of the top floor as a result of a large fire in September 1972. The attempts of the fire service are said to have been hindered then by the 'double-slating' of the roof – a wartime measure to prevent damage by incendiary device – and which doubtless led to greater damage than would otherwise have occurred. The original factory chimney has also gone, demolished as being unsafe. The main building of Yeo Valley Primary School was originally that of the Miller Institute, built by Alfred Miller for the benefit of his workers. Together with the factory it marks the furthest margin of the old Derby.

Beyond this the road changes its name from Vicarage Street to Derby Road and today continues into Gorwell Road and eventually out onto Goodleigh Road. However if you look to your left, just as you swing around into Gorwell estate at the top of the hill, you will see, between the red brick houses of Frankmarsh and the grey concrete of Gorwell, an old lane which seems to lead past the Kingdom Hall to nowhere. This in fact is the original Derby Road which is shown on Ordnance Survey maps as late as 1940 as continuing on up at least as far as the drive to Stoneyard Farm and was probably the old route to Goodleigh. Although now overgrown it still leads to Gorwell, or *Gor Hill*, House, the home which John Miller, the proprietor of the lace factory, had built for his family. The old gateway to the drive still survives and walking down the lane today it is not difficult to imagine him riding down the hill in his carriage to his factory standing proudly in the distance.

1

Rawleigh Cottages

EARLY BEGINNINGS

The history of machine-made lace in Barnstaple goes back to 1808 when John Heathcoat who was based in Nottingham and Derby patented the first machine capable of producing bobbin net lace.[1] Further improvements followed and some time in the next few years he opened a lace factory at Loughborough. In 1816 following an attempt to lower his workers' wages they rioted and destroyed the factory forcing John to relocate to Tiverton.[2] This new factory was successful and John established a London warehouse to facilitate sales. This was run by a John Boden who apparently had followed John from Loughborough.[3]

In 1822 Boden and John Heathcoat's elder brother Thomas decided to open a lace factory in Barnstaple locating their new works in a long derelict woollen factory at Rawleigh in the parish of Pilton.[4] The connection of Thomas and John with this undertaking is unclear but they seem to have been owners (or proprietors as they are described) who leased the factory out to managers. In 1823, for example, a local directory refers to Boden, Morley & Co. as lace manufacturers of Rowleigh [sic] Mill in Barnstaple.[5] This Morley was probably William Morley a Derby inventor responsible for creating the double locker rotary machine which was a notable advance in the field of machine made lace.[6]

The only details of this factory's earliest years are

1

contained in the 1822 reference quoted above which reads,

> *Barnstaple promises a resuscitation to its decayed splendour, as a manufacturing town. The mills and factory at Rawleigh, within about half a mile of the town, on a branch of the extensive estate of R.N.Incledon Esq., which for the last 20 years have laid almost useless and falling into decay, have been taken by some gentlemen who belong to the lace manufactory at Tiverton — and who are expending a very considerable sum of money in order to establish it upon the principle on which they have conducted, with so much deserved success, their large concern at Tiverton. This undertaking will prove highly advantageous to the town of Barnstaple and its vicinity, it being estimated that more than £40,000 a year will be circulated by the establishment.*

In November 1824 a small news item in the fledgling North Devon Journal reported that another lace machine was being put into operation in a building on the Quay in Barnstaple by a Mr.Simmons who, from a later reference, would seem to have been a foreman at the Rawleigh factory.[7] Two weeks later there is a reference to a case where eight workers at the Rawleigh lace factory run by Boden and Grace, were gaoled for absenting themselves from work.[8] This new partner was Robert Grace who must have replaced Morley as lessee sometime in the preceding year but no details survive about this change.

This new partnership didn't last very long as at some point within the next 10 months Boden dissolved it and purchased 10 acres of land and a house near Stoney Bridge, Barnstaple from Mrs. Abigail Holey of Bideford and her son James Hanmer Holt.[9] On this land he built a new lace factory which came to be called the Derby factory after the area from which he originated.[10]

A contemporary report in the Sherborne Mercury gave news of other developments as well,

> *In addition to the large lace concern established at Rawleigh, and the factory now building at Stonybridge for a similar purpose, a contract was concluded on*

Monday last, between Messrs.Symonds & Co. and the Corporation of Barnstaple, for a long lease of the marsh nearly opposite the Infirmary, on the Newport Road, for the purpose of building another Lace Manufactory.[11]

The Symonds mentioned here can presumably be identified as the Mr.Simmons who was erecting the lace machine on the Quay in 1824.

Thus within four years Barnstaple had developed from an ordinary rural market town to a hub of industry with three new factories mass producing textiles with the aid of the latest contemporary technology.

THE RAWLEIGH FACTORY

This the earliest of the Barnstaple lace factories had a short but lively history which can be pieced together by reference to newspaper articles and advertisements. Its origins have been charted above including the gaoling of 8 of its workers for absenteeism in 1824. In 1827 another 2 workers were taken to court for the same misdemeanour which would seem to indicate problems at the factory. The report on one of the cases reinforces this impression,

Mr John Heathcote, who appeared on the part of Messrs.Grace & Co., stated that they were compelled to make this example, in order to repress the spirit of insubordination, which was so dangerously prevalent among the junior classes of their work people.[12]

Perhaps one source of friction was the poor safety record of the factory. In November 1827 James Davey a young worker in the Rawleigh factory had to have his mangled arm amputated after getting it caught in the unguarded machinery.[13]

It is interesting to see John Heathcoat appearing in the court case as he was almost certainly based in Tiverton at this date. He wasn't the only non-resident proprietor/manager as is shown by a report concerned with the election of new burgesses (or voters) for Barnstaple in June 1830. At a crowded and noisy meeting

3

in the Town Hall the resident burgesses unanimously voted against the Council's choice *Mr.Grace lace factory owner*, on the grounds that he didn't live in Barnstaple.[14] A later article, however, lists Robert Weldon Grace as having been elected a burgess in June 1830 so the Council seem to have ignored their fellow townsmen.[15]

This animosity came to a head six months later when a wave of riots and strikes swept the country. Extra special constables were hastily sworn in at Barnstaple, Torrington, South Molton and Ilfracombe and public meetings were held to try and defuse the tense situation.[16] An attempted arson attack on a barn in Pilton was reported in the Journal which noted in passing the *well disposed character of the men employed at the Factory*. This was possibly wishful thinking as within a week the same newspaper was reporting, *We regret to state that no adjustment has yet taken place between the mechanics and proprietors of the Rawleigh Lace Factory: all the men remain out of work*.[17] This strike only lasted a few days but it is interesting to note that no unrest was reported at the other two lace factories which perhaps indicates underlying bad feeling between manager and workers.[18]

In this same year the ten year old daughter of a Goodleigh mason drowned in the lace factory's mill leat on the River Yeo as she *was carrying her home lace work to the factory at Rawleigh*.[19] This indication of the existence of child labour and outworking is one of the few references to how the industry was actually operating. Young labour was nothing unusual in this period prior to compulsory education and outworking was a common practice in other local industries, notably glovemaking. This tragedy didn't lead to any action as at least two other young children drowned at the same place.[20]

A year later the 27 year old son of Thomas Heathcoat described as *proprietor of the Rawleigh Lace Factory* drowned in the River Taw near the bridge at Tawstock.[21] Doubtless this personal tragedy hit Heathcoat hard but the lace business carried on, at least for two years. In May 1837 the Journal noted that *in consequence of the depressed condition of the manufacturing interest* all three lace factories in Barnstaple were closed.[22] Two of the factories, including the Rawleigh one, evidently re-opened fairly quickly but not before a public meeting was held to discuss the distress of the workers who had been laid off. At this Heathcoat appeared and noted that his labour force numbered 92 men with about the same number of boys and his weekly wage bill amounted to some £15 a week although when in full work it totalled £75. Whether this indicated a much larger workforce had been employed or that wages had been reduced isn't made clear. No indication is given of when the Rawleigh factory resumed work in the newspaper columns.

This temporary closure, however, along with the death of his son seem to have taken their toll of Thomas as he died on 18th April 1838 aged 63. His death was recorded at some length in the Journal,[23]

> *Mr.Heathcote had been in his usual health until the Wednesday before his death, when he was seized with sudden illness in his counting house, and was conveyed home; but under medical treatment he had pretty well recovered; on Tuesday night, however, he was seized with an apoplectic fit, and remained senseless until his death which occurred about one o'clock the following day. He was much respected as a man of the highest integrity, and has left a widow and nine children to deplore their bereavement.*

The ownership of the factory reverted to his brother John who seems to have taken an interest in both it and the workers employed there. In September 1840 for example a meeting of the Barnstaple Total Abstinence Society was held at the Rawleigh factory at the invitation of the *liberal proprietor.*[24] Heathcoat, who was also a member of the Society, had,

> *...fitted up and furnished it with with books and temperance publications for the mental and moral improvement of such of his workmen as are disposed to adopt the pledge, and has also provided for the instruction of those who are incapable of reading or writing.*

The attempt to educate ordinary factory workers was unusual, as at this date the only schooling most could expect came via Sunday Schools or primitive 'Dame' schools.

Unfortunately for Heathcoat, however, although he improved the working conditions of his labour force he couldn't improve the poor trading conditions of the lace industry. In June 1842 an article appeared in the Journal headed *Depression of Trade* which gives a clear insight into the problems assailing the lace business at this date. The factory referred to as likely to close almost certainly was the Rawleigh one.[25]

> *We have heard with the deepest pain that it is the intention of the respected proprietors of one, if not of both the lace manufactories of this place, to suspend the operation of a considerable portion of their machinery. We believe that this resolution has been arrived at with the greatest reluctance; but the immensely accumulated stocks of the article of lace, which it seems quite impossible to get off at any price, leave the proprietors no alternative. For years the trade has been known to be in a declining state, but we are told (and the Nottingham papers confirm the fact) that it is now unprecedentedly depressed. We cannot contemplate the prospect of fifty or a hundred families thrown out of employment in so small a population without the greatest anxiety; for the earnings of the lace operatives for years have been so small as to put out of the question the possibility of their making provision for an exigency of this kind; and to suspend their work is at once to doom them to the poor house. And looking at the dreadfully dull state of trade in this town, the means of help by way of public subscription must be limited, while any considerable increase of the already high poor rates will inflict serious injury on a class of payers who might more justly be pensioners upon the parish pay than contributors to it.*

It seems odd after this dire set of predictions that there is no follow-up in succeeding numbers of the Journal. Evidently the factory staggered on as in February 1843

16 year old George Puddington was standing at his machine in the Rawleigh works when the leather drive belt was dislodged from a pulley. As he placed it back his leg was caught up and his arm then caught in the machinery and severed above the elbow.[26] Incredibly George then ran home before being taken to hospital.

The problems identified in 1842 clearly did not go away as in June 1844 the newspaper published what turned out to be the obituary of the factory when it wrote,[27]

> We regret to find that there is a probability that the extensive manufactory at Rawleigh, in this borough, in which for many years past the lace trade has been carried on by Messrs Heathcoat and Co. will be shortly closed. Much of the machinery has been already removed to the premises of Mr.Heathcoat MP at Tiverton; and many of the hands have gone with the machines, and will be employed there. We fear, however, that some who have long found employment at Rawleigh will be wholly thrown out of work while the discontinuance of so large a concern cannot but have a considerable effect on the trade of the town, and especially at Pilton. We have not heard that the premises are yet engaged for any other kind of manufacture.

That the premises were left empty was shown by a court case in April 1845 when Samuel and Elizabeth Nott were tried for the theft of lead and brass from the Rawleigh factory which was described as unoccupied.[28]

For some reason it was another four years before John Heathcoat got around to advertising the building for sale. Possibly he had been looking for a new lessee but finding none decided to realise his assets; at all events the premises were advertised in the Journal in July 1849 along with thirty Mechanics' cottages. The advertisement gives some idea of how complex the undertaking must have been — speaking as it does of a main building 120 feet long on five floors along with separate offices, warehouses, finishing rooms, a smithy and carpenters' shop, a gas works, a Steam-warming Apparatus and two waterwheels.[29]

That it did not sell at once is evident from a small item in the newspaper two months later when Barnstaple Town Council applied to John Heathcoat for permission to use the derelict factory as a *house of refuge for those families whom it may be thought desirable to remove from infected and overcrowded dwellings* during an anticipated cholera oubreak. The MP *instantly and cheerfully complied* with this request although the buildings never had to be used for this purpose.[30]

The building actually reverted to its first use as a woollen mill a few years later and then was let again for other purposes in 1862.[31] From the foregoing it can be seen that the first lace factory in Barnstaple lasted just 22 years.

THE NEWPORT FACTORY

As previously noted the first reference to this factory came in 1825 when Messrs.Symonds & Co. took a lease of a site described as *the marsh nearly opposite the Infirmary, on the Newport Road* to build a *Lace Manufactory*.

The original proprietor was Henry Symonds who was described as *an enterprising and ingenious person who was employed at Rawleigh, and had a principal share in the superintendence of the machinery there.*[32] This factory was very modest being run by an 8 horsepower steam engine but it seems to have been the only one to have retailed its products locally the owner opening a *show room in Barnstaple, where worked lace dresses, veils, etc., were "finished" and retailed.*[33] Symonds went into partnership with Richard Sharland. There is an early record of a celebratory meal being taken at the factory to mark the marriage of Sharland who is described as *one of the proprietors.*[34]

Unfortunately for the two men this reference is about the only positive one that has been located to their venture. In February 1827 a young employee, John Phillips, was taken to court on a charge of absenteeism.[35] In August of the same year the construction of new lime

8

kilns at Coney Gut led to flooding of the surrounding marsh including the engine house of the factory.[36] Three months later Thomas Britton, a disaffected employee, was sentenced to 2 months gaol for absenteeism and *wilfully damaging part of a lace machine, called a spur*.[37] That this wasn't an isolated case was proved in June 1828 when a 17 year old apprentice James Darker was charged with being the ringleader of a group of *disorderly boys* in the factory.[38]

In August 1831 a *weak minded old man* smashed the factory's windows after being taunted by Symonds' employees.[39] Two years later Sharland took the landlord of the Globe public house in Barnstaple to court where the publican entered a recognizance of £50 not to attack the lace manufacturer.[40] In the same year Thomas Harvey was gaoled for stealing money and *a small instrument used by lace makers called a trick* from T. and W.Hill *who occupy apartments at H.Symons* [sic] *factory at Newport*.[41] This suggests that workers at the factory were possibly self-employed journeymen who hired working space in the factory which seems an odd arrangement but presumably had benefits for both groups involved.

This catalogue of problems continues via drunken workers, scrumping apprentices and accidents up until May 1837 when the Newport factory along with the other two in Barnstaple closed owing to *the state of the markets*.[42] Whereas the other two, however, weathered the depression and reopened soon afterwards Symonds and Sharland decided to close their business permanently and in December of that year the Journal carried an advertisement for a *Valuable Lace Manufactory In Bishop's Tawton, near Barnstaple, Devon... late in the occupation of Messrs.Symons & Sharland*.[43] In keeping with the unlucky history of the building even the advertisement was late in appearing owing to a mix-up in the newspaper office![44]

The building was not on the market very long as only three weeks later the Journal reported that,[45]

The lace factory, at Newport, near this town, with part of the machinery, was purchased at the public auction, on Monday last, by John Miller, Esq., of Gorrell House, the opulent lessee of Stony Bridge factory.

Whether Miller ever ran the factory as a going concern is open to doubt, certainly no references to its operation following his purchase of it have been located. In May 1845 a Town Council meeting heard from the Mayor that,

Mr.May, on behalf of the family of the late Mr.Miller (proprietor of the Derby lace manufactory) had proposed to *purchase a portion of Gooseley-marsh, on which he would erect two genteel dwelling houses, and destroy the factory at present standing there.*[46]

This was reckoned to be *a great improvement to that part of the town* which suggests the old factory had become an eyesore. The council, however, rejected the application, probably because the land was earmarked as a possible site for the rail terminus Barnstaple hoped to develop.

Clearly the building was demolished at some date and it is probable that its remains are under present day Rock Park. A sad end for this ill-starred factory which had an even shorter life-span than the Rawleigh business.

That Henry Symonds didn't entirely withdraw from the lace industry is shown by a notice in the London Gazette in 1841 regarding the dissolution of the partnership between Michael Snell and Henry Symons [sic] who described themselves as dealers in wine, spirits, ale, porter and coal as well as being *Lace Spriggers.*[47] This odd occupation refers to those who dealt in small pieces of lace that were later made up into large composite pieces.

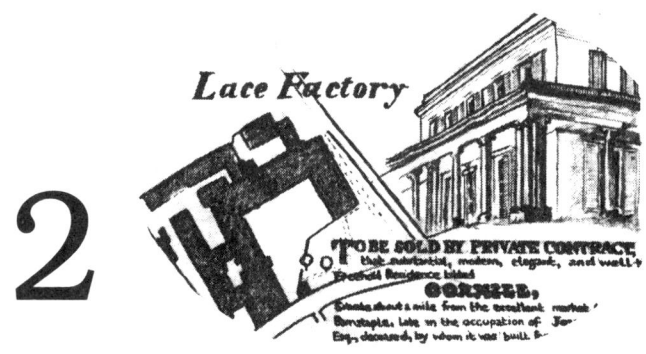

2

Gorwell House and map of Derby Lace Factory

THE DERBY FACTORY

As has already been explained this factory was built from scratch in 1825 by John Boden on a 10 acre site known locally under a variety of names – Stoneybridge, Brickfield or Brickyard. The factory is actually recorded under the first of these names rather than the usual one for some years. No details of its construction phase other than a report of an accident to one of its builders survives.[1] Its site was some way out of the existing built up area of Barnstaple but this isolation rapidly disappeared.

In December 1825 whilst the factory itself was being built it was reported how speculators had rapidly moved in to develop housing for the workers.[2]

A field adjoining the town of Barnstaple, containing something less than an acre, which the late proprietor purchased twelve years ago for £70 was sold about a fortnight since for £550. The field adjoining which contains two acres and was purchased by the father of the present seller for £140 was last week sold at the astonishing price of £2000!! – It will readily be concluded that these purchases have been made for the purpose of building: the latter is intended for the site of a complete street from Stoney Bridge to Ebberly Place, and afford ample accommodation for the increased population which the new lace factory must necessarily collect. Upwards of 120 houses are already engaged.

11

The next week the same newspaper reported the laying of the foundation stone of *the 3 new streets, intended to be built in Vicarage Meadow, at Barnstaple.*[3] The first of these was called Union Street and the joint developers were named as Benjamin Baller and William Thorne. According to one writer Boden himself built a row of houses in 1827 known initially as Boden's Row to house his workers.[4] This later became Corser Street following the marriage of Boden's daughter to the Rev.John Corser. Princes Street and Reform Street followed, the latter in 1832 being named after the great Reform Act passed by Parliament in that year.

The factory seems to have run fairly smoothly having just a few absentees so it is perhaps surprising that Boden sold out his interest in the business in 1828.[5] The handover took place on the 15th. of May in that year and was recorded in the Journal which gives a valuable insight into the size of the undertaking.[6]

> *On Thursday last, being the day on which a change of proprietors took place at the Stoneybridge lace manufactory, the whole of the hands employed on the establishment, with the wives of the same, amounting altogether to upwards of five hundred persons, were treated with a holiday, and a plentiful supply of liquor and victuals; the females, preferring tea, were accommodated therewith. The greatest order prevailed during the whole time occupied by the regale, and the various persons on the ground demonstrated their gratitude to their old employers Messrs Boden, and their confidence in their new, Messrs Miller & Co by drinking their health in copious libations.*

John Boden evidently went back to manage his far larger premises at Castlefield, Derby where, by 1831, his factory was the largest cotton plain net maker in Britain.[7] Interestingly the Boden group returned to the South West in 1892 when they purchased a factory in Chard.

John Miller, the new owner, came from Loughborough and presumably had some working relationship with John Heathcoat who constructed his first factory there.[8] A notice re the dissolution in late 1828 of a partnership between Miller *late of Loughborough* but *now of Barnstaple* and John Oram of Chard, lace manufacturers, exists.[9] This also indicates that Miller had interests in Nottingham as well though the details are not given.

Unlike the other two factories the Derby one appears relatively seldom in the newspapers during its early years though December 1831 saw the announcement of the death of James Street foreman at Miller's factory aged just 36.[10] Intriguingly one of his executors was a Henry Symons, presumably the owner of the Newport factory which suggests close connections between the men running the Barnstaple lace industry at this date. In May 1832 a happier announcement appeared about the birth of a daughter to the unnamed *lady of J.Miller Esq of the Factory.*[11]

This coincided with the passing of the Reform Act which was the first attempt to clean up the notorious state that politics in Britain had descended to with its bribery, rotten boroughs and nepotism. Miller evidently supported it as he allowed his employees to march in a great procession through Barnstaple to celebrate the Act whilst carrying a flag made of lace bearing the motto 'Union and Reform'.[12] This taken with the names of the streets housing his workers suggests that Miller was an unusually reform-minded factory owner. Such a radical stance was often associated with non-conformist religious beliefs and Miller is recorded as the builder of a Baptist chapel in Boutport Street at a cost of £900 which is still in being today although the original building was completely rebuilt in 1860.[13] Two years after its opening the Reverend Richard May, who was Miller's brother-in-law, was appointed minister there.[14]

That his reformist views were shared by his workers is shown by the case which came to the Petty Sessions court concerning two of his employees who got into a fight with a village constable over 'Reform' at a public house in Bishops Tawton.[15]

Unfortunately for Miller his men became rather too enamoured of radical politics and went on strike in November 1833. The report in the Journal gives some fascinating details of working hours which suggests that Miller got full worth from his workers.[16]

On Monday last, the operatives in Mr.Miller's lace factory in this town, simultaneously quitted their work; the cause we understand to have arisen from a dissatisfaction on the part of the men to the length of time they were required to labour, being from five o'clock in the morning to seven at night, with intervals of half an hour at breakfast, and three quarters of an hour at dinner; and, as must be necessary in a concern where such numbers are employed, the time is marked with great exactness, and when either of the men were a few minutes too late, the gates were shut against them, and they were mulct of a portion of their wages; and according to their representation many of them live at a considerable distance from the mill, and cannot go to their homes, get their breakfast, and return to their scene of labour within the half hour, they therefore deputed some members of their body to confer with the master, in order to obtain an extension of their time to an hour each, at breakfast and dinner, to which Mr.Miller returned a negative, but offered to extend the time for breakfast to three quarters of an hour; but this would not satisfy the men, and they in consequence struck, and still continue out of work.

The following week the Journal coupled news that the strike was continuing with a denunciation of those who had sent anonymous threatening letters to the foremen at the factory who were still working.[17] This nasty turn of events became even worse to judge from a long report in the next issue of the newspaper which indicates that Miller employed 'blacklegs' to take the place of striking operatives. The coverage sprang from a fight between a striker called Nott and two 'blacklegs' called Horne and Scamp at the Golden Ball public house in Barnstaple.[18]

The case came up before the Mayor who was far more interested in the landlord William Symons and what went on at his premises. In answer to questions from the Mayor Symons agreed that, *there were several societies which held their meetings at his house* amongst them a

Trades' Union whose object was merely *to cultivate friend-ship*. It had been established some 4 months, had a range of officers and met at least once a week. All this seemed fairly innocuous but the landlord then added, *there is a guard who stands masked at the door, during meeting hours, holding a drawn sword, the society adopts this precaution to prevent the intrusion of any spies, who might report the proceedings to the master; it is a secret meeting.*

Symons was himself a member having joined at Tiverton which suggests two things, firstly that the Union was well organised in Devon and secondly that laceworkers moved fairly freely between the lace factories in the county.

Miller was the next witness and he was insistent that Union members took an oath – *and that it is because of the peculiarly solemn nature of that oath, that many of them, who would be very glad to come back, cannot return to my employ.* He further claimed that the oath enforced solidarity in that no workers would return unless he took them all back as a body.

The Mayor then returned to Symons asking about this oath but the landlord refused to answer directly and further refused to give the name of the Union secretary even though threatened with 3 months imprisonment. On a more conciliatory note he was sure that the strikers would accept arbitration in their dispute with Miller if the latter agreed. The millowner, however, responded that the strikers had refused to talk to him when he approached them and in any case he was *receiving fresh men into my employ every hour* and thus couldn't be troubled with the strikers adding, *I will never employ any of them until they leave the union; and unless they return speedily, none of them will be wanted.*

Evidence was then heard about 100 strikers insulting and hurling mud at 'blacklegs'. One of the latter, James Priscott, was a lapsed Union member who gave details of his initiation into the organisation. On entering the room at first his eyes were *bunked* and he was led around the room, told to kneel and then rose to be shown an apparently human skeleton before taking the oath of membership. At the end of the hearing three women

15

involved in Priscott's case were fined as was the attacker in the original assault case.

The Mayor evidently stepped in to mediate between Miller and his employees as the following week the Journal editor hailed the return to work of the strikers adding that *a number of new hands have been engaged, which must necessarily exclude several of the old ones.*[19]

Unfortunately the editor was rather too hasty and in the next week's paper is a long report of the continuing stoppage. This was due to Miller having taken on 40 to 50 new workers and refusing to re-employ 30 of those who had struck. At 4 a.m. on Friday the 29th. November *the unemployed hands with the operatives at the other two factories, congregated around the gates, to prevent the employed labourers from entering the factory.*[20] The Mayor turned up at 9 a.m. and tried to persuade the crowd to disperse but to no avail though he eventually got every-one to agree to discuss their grievances in the Guildhall the next day when the Deputy Lord Lieutenant of Devon, Lord Ebrington, and the local MP J.Chichester would attend.

The meeting began with the Mayor warning some unnamed individuals *not at all concerned in the business* to stop inciting the peoples' feelings. He then gave way to Lord Ebrington who agreed that the laceworkers had a right *to combine and consult together for the purpose of fixing the terms on which you will engage yourselves, but you have no right to offer any intimidation to other parties who may be willing to work on different terms from those which you may be disposed to accept.*

Nine *delegates* from the workers then entered the jury box and Miller put his side of the case. Apparently there were only 8 men who had not been re-employed and he offered to take them back so long as they agreed to leave the Union and work on whatever machine they were assigned to. This he was willing to do even though the trade was *in a most deplorable state* only being kept going by *the American demand.* He then challenged the men to identify any times when he had acted unjustly towards them. Henry Davis, one of the *delegates*, pointed out that the strike began after Miller refused to grant the men

16

extra time for meals *as the other masters allowed.* Another worker, Thomas Glass, accused Miller of cutting 4d from the wages of men who were only a few minutes late which was roughly a fifth of their daily earnings and seemed harsh. The mill owner prevaricated over this charge which led to hissing from the crowd in the Guildhall and a promise by Miller to *submit my books to a committee* for impartial examination. He then returned to the poor state of trade and wage rates which he claimed were *greater than at any other place this side of Tiverton* (a claim that was challenged) even though lace which had sold in 1828 for 18d per 'rack' now only fetched $7^1/2$d. Miller admitted that pay had been reduced per 'rack' but because of heavy investment in new machinery since he had taken over the men were earning just as much as before.

He then produced a document he expected all those wishing to be re-employed to sign. This was read out and a clause about agreeing to leave the Union was greeted with shouts of *No! no! never!!* from the workers in the room. Henry Davis added *Sooner than violate my oath, I will go with my hat in hand and beg my bread.* Miller then offered to take back any employee who turned up on Monday and signed the document – and with this the meeting dispersed.

Come the Monday a large posse of special constables accompanied the Mayor to the factory gates but there was no disturbance. The Journal noted that,

> *The staff of the North Devon Militia were also under arms, in readiness to render their assistance.*

A later report noted that only 4 or 5 of the strikers had not returned to work. These exciting events were finally drawn to a close a few weeks later when the Mayor received a letter from the *principal secretary of state* (the Prime Minister) Lord Melbourne praising his actions during the dispute.[21]

Over the next few years the factory and its workers carried on far more peacefully. In October 1835, however, the factory was broken into and 31 machines on the large ground floor room had the lace being prepared on them cut and slashed to pieces.[22] The Journal report stated that,

No suspicion attaches to any of the hands employed upon the premises, between whom and the master the best feeling prevails.

In fact, nobody was arrested even though Miller offered the then huge reward of £50 and we are left to wonder if this was the work of a disaffected employee or a North Devon 'Luddite'?

Three months later the new Barnstaple Town Council met for the first time following the elections brought about by the Municipal Corporations Act of the previous year and Miller, already a member of the old corporation, was elected an alderman by his fellow councillors, a clear indication of his importance to the life of the town at this date.[23] He probably used his position to help divert a proposed new Workhouse from being sited next to his factory, a move supported by an editorial in the Journal which saw dangers in having the two buildings next door to each other.[24] A few months later he was querying the imposition of quay dues and although he did not declare any personal interest it is clear he was seeking lower charges for his own advantage.[25]

This is one of the few references to a central part of the lace industry in North Devon – the importation by sea and later by rail of cotton and silk thread. The former came from the Lancashire mills and the latter from Italy, France, China and India via merchants probably based in London.[26] The organised and cosmopolitan nature of the lace industry at this date is shown when in November 1836 Mr.Miller was absent on business in Nottingham when the council voted on the appointment of a new magistrate.[27]

Even though he had become an alderman Miller wasn't an establishment man as was shown by items in the Journal in April and May 1837. At this period non-conformists were campaigning against the compulsory payment of church rates which were used for the upkeep of Church of England property. At one public meeting about this issue Miller, a strong nonconformist, had allowed his workers to attend.[28] He is reported as saying,

I am a dissenter, and I feel that the dissenters have been grossly maligned...today.

He went on to deny that dissenters only did business with others of their faith – a statement that was cheered by the crowd present. When a vote was taken Miller's view won the day although boys from the factory were thought to have voted even though it was forbidden by the chairman.

A letter in the Journal the following week defended Miller's views and turned on his attackers from the Church of England alleging they had only managed to gather supporters by distributing free alcohol![29]

All these disputes, however, were eclipsed by the closure of all three Barnstaple lace factories in May of that year. This has been previously noted in the sections on the other two factories but by this date the Derby factory was by far the largest and the following Journal report should be read with this in mind,[30]

We exceedingly regret to state that in consequence of the depressed condition of the manufacturing interest, the lace factories of this town are all still. The suspension took effect on Friday last. It is painful to see so many industrious mechanics strolling about the streets unemployed, and it is yet more painful to know that in the majority of instances, now that they have no labour they are without any resources for the sustenance of themselves and families: nor is this the result of their own improvidence; for such has been the state of the markets that for a long time past their wages have been but barely sufficient to meet their daily wants.

The seriousness of the situation led to a letter from Thomas Hill (described in a later note as a lace manufacturer of Nottingham but late of Barnstaple – possibly the T.Hill who had some tools stolen from the Newport lace factory?) being published in which he reckoned that there were *800 men, women and children, employed by the factories* in the town who circulated at least £200 per week of *ready money* amongst the *shopkeepers, farmers and various other trades* of the area. Interestingly he refuted the widespread belief that the problems in the lace industry were due to the product becoming unfashionable saying that it was the economic decline in the USA, which was the largest customer, that had created the situation mill owners now found themselves facing.[31]

These 800 workers had to apply to the parish authorities for relief, i.e. the dole, and this sudden immense burden on the town's resources led to a public meeting being called in the Guildhall though by the time it was held the Derby factory had resumed working. As Miller was away on business *in the north* his clerk a Mr.White spoke for him saying that *he did not believe that there was much distress among the men* though one of them then got up to claim that his wages *for some weeks past* had *not averaged more than 3s.*[32] White offered to open his employer's books to *ascertain what all the men had been earning for three, six, or nine months*. He added that, *He was particularly anxious to correct an idea that there was any probability that the works would be altogether stopped, for, so far as the Derby factory was concerned nothing could be more fallacious.* The meeting then agreed to look into the possibility of providing special relief to the laceworkers but in the event this was refused though it should be added that both the two main factories began full work fairly quickly after the stoppage.[33]

Evidently at this time the North Devon Advertiser, a local Tory-leaning paper, had published a letter from one of Miller's workers called Colwell who claimed to have only earnt 4/- per week for the last fortnight out of which he had to pay a boy who assisted him at his machine.[34] Presumably the Tory paper didn't much care for Miller's liberal views but the mill owner published a letter in the more sympathetic Journal refuting this allegation of low pay setting out Colwell's earnings. This reads,

20

	He received			his daughter			his wife		
	£	s	d	£	s	d	£	s	d
April 6	0	16	11		2	6		8	3
April 13	1	0	6		2	1		9	6
April 20	0	9	9		2	2		6	8
April 27	1	2	7		2	0		9	5
	3	9	9		8	9	1	13	10 -£5 12 4

Over these four weeks the wife paid 16/- *for assistance* and Colwell paid his boy assistant 15/- *leaving upward of £1 a week for the maintenance of the family.* Such wages at this date were fairly generous when it is considered that the contemporary farm labourer's average weekly wage in the area was 10/-.

Doubtless Miller's actions two months later enraged the Advertiser's editor and supporters. For some time, as noted previously, there had been an argument over the payment of church rates by non-conformists. Evidently Miller had decided to take a stand on the issue as in August 1837 he was summoned to court for non-payment of some £3.13.4 in unpaid church rates.[35] At the court hearing he read out a long statement the core of which was summed up in one sentence,

I object to church rates, because I consider no man ought to be compelled to support any system of religion. "God requireth a cheerful giver", and nothing but what is freely given is acceptable to him.

He went on to attack the whole idea of child baptism ending (rather lamely),

Having thus protested, I shall pay the rate when legally demanded, but I shall resist the making any future rate, if I know of one being proposed.

This is one indicator of his beliefs, another came a few months later when, in a move copied later by Heathcoat at the Rawleigh works Miller *gave those of his men who are members of the Temperance Society (to the number of upwards of 30) a plentiful treat of fruit cake and coffee.*[36] On the following day Mrs.Miller *gave a similar treat to 75 children belonging to a charity school, at Derby.*

Nonconformism was often associated with the temperance movement and teetotal workers were more trustworthy than those who drank.

In November 1838 Miller was re-elected as an alderman on the Town Council[37] whilst in early 1840 his liberal views came to the fore again when he appeared at a public meeting in Barnstaple held to discuss the repeal of the Corn Laws then in force.[38] Miller actually proposed the successful motion arguing that his workers needed cheaper food. He proudly stated that his employees *had been kept in constant employ; their wages had been enough to find them food; certainly they were not what he wished them to have, but still they were better off than artisans in many other places.* In an aside he added *his state of health would not permit him to go at length into the question* – the first hint of the illness that was to kill him.

In a strongly contemporary sounding report a few months later a councillor complained of pollution from the Barnstaple gas works which was producing noxious smells and run-off that was killing fish in the river. A similar nuisance was said to come from *the draining of the gas works of the Derby factory into the mill leat.*[39] Miller doesn't seem to have done anything about this, possibly because he was experiencing more symptoms of his fatal illness. There is further evidence of this from November 1841 when the Town Council visited Lord Ebrington at his seat at Castle Hill, Filleigh. A small item in the newspaper noted that Alderman Miller was too ill to attend.[40]

His illness, however, didn't stop him from discussing his nonconformism at a public meeting called to send congratulations to the Queen on the birth of an heir in the same month.[41] There was an argument over how the meeting was called and *Alderman Miller remarked that time was when to be a dissenter was to be considered disloyal; but he had thought that day long gone by.* Only two weeks later the birth of his ninth child was recorded.[42] The birth was not an occasion for rejoicing, however, as within 3 hours his wife Jane was dead. Described as a lady of *high moral and religious principle* she was aged just 40.

This blow was followed a few weeks later by a daring robbery at the Derby factory when a burglar, though disturbed before he could really steal much, managed to get away with £6.5.0 in cash.[43] The police soon arrested 17 year old James Nutt *belonging to a notorious family in this town*. Taken to police cells he managed to escape within a few hours, was recaptured but then released as the evidence against him was only circumstantial.[44] Miller, calling on help from his fellow members of the Barnstaple Association for the Protection of Property, had Nutt brought to trial where he was found guilty and sentenced to transportation to Australia for 7 years.[45]

As if the death of his wife and this break-in weren't enough, in June of the same year the Journal carried a report of the *unprecedentedly depressed* nature of the lace industry in Britain which would probably lead to the closure of the two lace factories still operating in Barnstaple.[46] The news item ended,

> *Gloomy indeed is the condition and prospect of the country generally in its commercial and manufacturing interests; and, although principally an agricultural dis-trict, we appear to partake our full share of the prevail-ing depression and despondency.*

Within four months John Miller was dead.[47] His obituary reads,

> *At Gorwell house, in this parish, on Thursday night last, John Miller, Esq., proprietor of the Derby lace manufactory, and an alderman of this borough, aged 55. Deceased had suffered several years most distressingly from a disease which a post mortem examination discovered to be a cancer in the head. From his extensive business and the employment it supplied, Mr.Miller possessed very considerable local influence; and his death will be deeply deplored by a large circle which the amenity of his disposition had attached to him, and by his numerous dependents; but principally will the bereavement be felt by his interesting family of nine children, all of tender years, who have been called to sustain the loss of both their*

natural guardians within eleven months. His friends are consoled by the assurance that his protracted affliction had matured his virtues and piety, and that, released from a state in which it is the lot of but few to suffer so much, he is now "ever with the Lord". His interment took place this day; and many of the Town Council evinced their respect to the memory of their deceased colleague by following his remains to the grave.

Two months later there was a sale of the late millowner's two horses, carriage and phaeton (a light 4 wheeled carriage) with potential buyers being directed to John May, Miller's brother-in-law, at the Factory.[48] This was followed by the sale of 'Gorhill' along with 25 acres of ground; the advertisement noting that John Miller had built the house for himself and his family.[49] The buyer would appear to have been the Rev. Richard May, the Baptist minister referred to earlier who was the brother of John May. The factory and business had been taken into hand by trustees appointed under Miller's will. They included Edward (or Richard) Harris, Frederick Maunder, a prominent woollen merchant, John Willis, Rev. Richard May and Thomas Hill of Nottingham.[50] The report which gave these details added that they were supervised by William Hill of Nottingham and John May, the latter of whom was running the factory on a day-to-day basis.

Miller's will also specified that his two eldest sons John May Miller and William Walter Miller, should take over the running of the business once all the siblings had reached 21 years of age whilst he left £2500 to each of his four boys and £2000 to each of his five girls when they reached adulthood.[51]

Miller's affairs seem to have taken some time to sort out as it wasn't until May 1845 that a request was inserted into the London Gazette for his creditors to apply to the Court of Chancery for payment *in the cause Miller vs.Harris.*[52] This was repeated nearly two years later although whether this was a reflection of the fabled slowness of Chancery actions or of the tangled state of Miller's finances isn't clear.[53]

The next news of the factory rather than the workers

came in September 1848 when Richard Thorne, a carter, was fined 2/6 for driving his cart dangerously when employed to haul coal from Barnstaple Quay to fuel the boilers at the Derby Lace works.[54]

A few weeks after this Miller's *eldest surviving daughter* Harriet Charlotte died at Exeter aged 32 although whether she was married or not is unclear from the announcement.[55]

John May who, as well as acting as manager, had also been a town councillor for some years, lost his seat, along with a prominent ex-Mayor, in November 1849.[56] Only a few weeks later he made the news again when he and *Mr.Miller* discovered a cache of stolen goods whilst walking through fields near to Barnstaple.[57] This Miller was probably John May Miller, John Miller's eldest son, who *came of age* in February 1850 and became more involved with the running of the business. The Journal covered this event in some detail noting in passing that the wages bill of the factory was then running at £5000 per year.[58]

Although the factory was still under the control of the trustees John May Miller was beginning to make his mark. In June 1851, for example, he appears as a trustee of the then newly-formed Barnstaple and North Devon Permanent Building Society involved in the sale of a house in Princes Street.[59] The following year he is reported as giving an employee £5 when she emigrated to the USA[60] and in the same year another more elderly employee was retired with a pension – an event unusual enough at that date to be thought worthy of coverage in the newspaper.[61] Two years later Miller was recorded as giving coal to the factory employees during a particularly cold period in January of that year.[62]

That John wasn't content just to see his father's business ticking over but had a real interest in developing the factory is evident from two news items in early 1854. In the first it was noted that business had resumed at the factory *after a suspension of about ten days, during which a new engine, of greater power than its predecessor, has been erected. While the hands have been unemployed, their families have been the objects of Mr.Miller's solicitude, and he has generously ministered to*

25

their necessities.[63] The second was a report on a celebration dinner held at the Newington Inn for 40-50 of the factory workers to *celebrate the recent erection of the new and powerful engine.*[64] The guest speaker Thomas Seldon senior exhorted the men *to persevere in a course of industry, sobriety, and economy, that they might in the decline of life enjoy a state of ease and comparative independence* whilst the evening finished with toasts being drunk to *the health of the spirited proprietor of the factory.*

This settled if paternalistic view of the undertaking was effectively destroyed in January 1856 when John May, described as manager of the Derby lace factory, appeared in court at Exeter to file for bankruptcy.[65] The main witness was John May Miller whose cross-examination provides a wealth of detail about the factory in particular and industry in general. May was being paid £250 per annum as full-time manager up until February 1850 when, with the consent of John Miller's executors he was allowed to draw his salary *without devoting the whole of his time* to the business. Presumably John May Miller was becoming more confident of his ability to run the business on his own. With time to spend on other interests May took over the running of a mill owned by Mr. Maunder in Barnstaple (possibly the old Rawleigh factory) which was involved in the *doubling business.*

May had an agreement with Miller that *he should be supplied with cops* (conical balls of thread on a spindle) from the Derby factory and this guaranteed source of work led to him spending £800 on buying second-hand machines at Manchester which, apparently, were *a great bargain.* The business seems to have never been too prosperous and in October 1855 when May became bankrupt he blamed his failure on *want of capital and insufficient water power until July last to work the machinery profitably.* His debts came to £3827.1.10 although much of this was made up of mortgages on property which were covered by the value of the property itself. Indeed the difference between credit and debit amounts was just £1.18.0.

Although the newspaper report on these proceedings is hard to follow it would seem May treated the finances of the Derby factory and his own as being interchangeable. At one point, for example, it was noted that he *had continued, since March 1852, to be the receiver of the rents of the freehold and leasehold estates of the late Mr.Miller, but had passed no account of his receipts and payments during the last three years, and how he stood on that account they had no means of knowing.* Again he *was in the habit of drawing from the Derby establishment cash on account, without any settlement.* It is all the more surprising, therefore, that two of his principal creditors, William Avery and J. Tatham, had appeared at Court and *asked the Commissioner to grant the bankrupt a first-class certificate, as his conduct had been fair and upright.* Can one detect an example of early Victorian class-solidarity here? May was, in fact, granted his discharge and seems to have returned to managing the Derby works full-time although one suspects under a tighter rein than previously.

In April 1857 John May Miller was elected to the post of churchwarden of St. Mary Magdalene presumably having become a member of the Church of England.[66] Although not following his father in religious matters he did follow in his father's footsteps seven months later when he was elected to represent the North Ward of Barnstaple on the Town Council.[67] Miller clearly took his civic duties seriously and his name appears in the Journal's monthly reports on the discussions of the council with some regularity and he soon became Chairman of the local Liberal party and, as will be detailed later, Mayor of his home town.

In 1859 he gave a dinner for his employees. Held on New Year's Eve *the liberal proprietor* as he was termed in the Journal's report *entertained those in his employ to a substantial dinner, and their wives and daughters to tea* the meal being followed by *a ball in the Infant School Room.*[68]

In February 1860 he was nominated as an ensign in the First Company of the newly established Barnstaple Rifle Volunteers.[69] These part-time soldiers were recruited on a local basis and for many the attraction seems to have been the social side of the organisation rather than serious soldiering. All the officers were, naturally, drawn from the upper ranks of society and Miller became a Captain when the Second Company was formed in September 1860.[70] Within a few years, however, he was resigning his commission claiming ill health.[71]

A far more prestigious and important post came his way in 1860 when he was appointed as a magistrate to the Barnstaple bench.[72] In the early days of the professional police force such magistrates took a far more active part in dispensing justice and monitoring law and order than we expect today. Being a Justice of the Peace was to be a real power in the community.

It seems rather odd, therefore, as a pillar of the community, that within a month he was appealing against an increase in the rates (from £167 to £300) levied on the lace factory,[73] although he was, however, just one of many who were angry at a general rise imposed by a group of new assessors. When the case came to be adjudged all the ratings were listed and it could be seen that Miller was the joint highest rate payer along with the Barnstaple Water Company.[74] The new rates were found to be far too high and the assessment was quashed.

3

View from the former Miller Institute

THE MILLER SONS TAKE OVER

The next important date in the history of the lace industry came in 1863 when the third of John Miller's three sons who were to run the business in future came of age.[1] John May Miller, William Walter Miller and Alfred H. Miller took over the complete operation of the business from the trustees of their father's will on the 15th. July of that year. A fourth son, Dr. R. Miller, didn't enter into the business at all. A fifth son, named George, is a shadowy figure in the family's history.

Of the three William took over the running of the wholesale department of the business which was based in Nottingham where he both lived and died.[2] He married twice, first to a Miss Gribble, daughter of a prominent Barnstaple banker, and second to a Miss Loviband of Bridgwater and had three children, dying aged 71 in 1901. Alfred, who was born in 1837, joined his brother John in running the Derby factory after initial plans to become a doctor were thwarted by an accident.[3] Described in his obituary as *A man of exceptional culture* he was greatly interested in books and was a fervent supporter of the North Devon Athenaeum. He seems to have been of a rather retiring nature only ever standing once for the Town Council (as a Conservative) and then *was not sorry he was defeated*. He did become a magistrate but was more noted as a keen shooter and fisherman. He

29

died unmarried in 1903 whilst on a trip to Nottingham.

The actual occasion of the brothers taking over the business was noted by a very fulsome tribute in the Journal which, even allowing for journalistic hyperbole, gives an idea of their characters. It reads,[4]

> *We congratulate the town on the fact, that the gentlemen who have so spiritedly and satisfactorily conducted the business for 12 years past will continue connected therewith as proprietors – free from leading strings or the control of trustees. To the employees, the announcement will be peculiarly welcome; as they have now the prospect of working under the direction of masters who have, by their liberality and kindness of heart entitled themselves to the confidence and esteem of all with whom they have come into daily contact, and whose integrity and honour have established for them a high reputation in the commercial world.*

Four months later John May Miller was elected unopposed as Mayor of Barnstaple.[5] His inaugural dinner was held at the Royal and Fortescue Hotel where he was again showered with tributes this time from his fellow councillors.

Although the brothers had got control of the family factory and achieved great municipal success, their business was apparently suffering a serious decline at this period. Speaking in 1872 John May Miller referred to the problems that began in 1860 with the outbreak of the American Civil War. Up until then trade had been good with much of the product being exported to the USA but with the war *every branch of textile manufacture was paralysed.* He referred to the machinery in their factory which *was idle from '60 to '64* and to many (of the workers) having to seek employment elsewhere, and said he *thanked God heartily that times were different now and that machinery was working which had long been idle.*[6] Presumably the unused machines did not produce the whole of the factory's output as the business certainly continued to function over the 1860s.

In 1868 John May Miller was elected as an alderman of Barnstaple, a position he occupied *very many years.*[7] At some point he also became a long-serving member of both the Barnstaple Bridge Trust and the Taw and Torridge Fishery Board.[8] A small glimpse of the charitable nature of John and Alfred came in 1871 when five of the boys employed in the factory were taken to court for absenting themselves from work on the day of a local regatta.[9] Only two of the boys were actually charged and then *not out of any feeling of vindictiveness*. Indeed the only reason was that if *they absented themselves they caused their employers serious loss, for they prepared the work which the men did, and unless it was prepared the men were necessarily kept idle*. The two boys were dismissed with a warning not to misbehave in future.

That the brothers were enlightened employers is reinforced by an offer they made to their employees in 1872. In February of that year they offered their workers two new wage structures.[10] In the first, working hours would be reduced from 61 to 54 per week but wages would be kept at the existing level. In the second, meal times would be increased from the $1/2$ hour for breakfast and $3/4$ hour for dinner to 1 hour each and the wages again maintained (with extra pay for the odd $2^1/2$ hours). The men agreed to the latter whilst those machine operators who were on piece work all had their wages increased by between 10 and 12%.

Following this generosity the employees held a dinner at the Barnstaple Assembly Rooms where John May Miller was the guest of honour (Alfred was unwell and William was in Nottingham).[11] At this event John was presented with an illuminated address from his men (women do not seem to have been invited) and many toasts were drunk. It was during his speech of thanks that John spoke of the problems experienced in the early 1860s. He also added the fact that he had spent 28 years associated with the factory in Barnstaple apart from 4 years he had been working in Nottingham – which suggests he had experienced all the intricacies of the lace trade at first hand.

Nine months later Allen Trist, the factory foreman, retired and the Miller brothers gave him a special presentation owing to his long service.[12] Described as a *noble type of the genuine working man* he was given an address on vellum and a large walking stick with an ivory handle and silver ferrule. He died, apparently still in harness, in 1881 aged 72 having worked at the factory for 50 years.[13]

The following year John May Miller in his role as alderman became a Guardian of the Poor i.e. one of those charged with running the Poor Law, that system of support for the poverty stricken built around the Workhouse.[14] The month of June 1873 saw the first of a long running and popular series of day excursions, called a *Wayzgoose*, organised by the Millers for their workers. This first trip was to Torquay and 50 employees went there by train and had a free dinner provided by their employers.[15]

Exciting news came in April 1874 when the Journal announced that,[16]

> ...*so good has the trade of the Lace Factory in this borough been of late, that the proprietors are about adding an additional wing to the building for the use of a larger class of machinery. Large excavations are being made for the foundations, and scores, if not hundreds of tons of soil have been hauled the last few days from the spot to the works of the Ilfracombe railway at the Quay.*

This wing was to the right of the main entrance gate. Surprisingly perhaps the paper has no other comment on this enlargement of Barnstaple's main factory.

This happy, if paternalistic, state of affairs did not last. In September 1874 the Journal printed a long piece headed *Lock-out at the Derby Lace Manufactory*.[17] Tracing the trouble back some 18 months previously the dispute had grown from the workers' wish to join a national Trade Union – a move opposed by the Millers. Intriguingly the Derby factory was one of only three factories where the Nottingham-based Union was not represented, the other two being Boden's at Derby and Heathcoat's at Tiverton – presumably the North Devon connection forged in the 1820s was still in operation. The Barnstaple men had been convinced of the need to join the Union after a visit from some Nottingham lace workers who explained the advantages of membership and pledged support if they decided to challenge the Millers.

When the Millers heard what was happening they *saw all their men and gave them all a week's notice to expire, on that day week, unless they in the meantime would withdraw from the Union.* The lace workers from Chard where the Union was well established came to Barnstaple to try and change the employers' minds but with no success. At the expiration of the week the 34 men involved were locked out and their absence led to the laying off of another 150 boys and women employed in the factory.

The same report reckoned the confrontation followed the unpopular sacking of three men two months previously but this was only conjecture. What was more certain was why the Millers were so opposed to the Union. Apparently it had *arbitary rules* about *the hours of work* and *the wages paid* which didn't fit with their views.

The report also gave details of the state of the lace trade at this date. The previous three years had been very prosperous but from 1860 to 1869 it had been very bad – *so bad that, for a long time together, every week's work involved considerable loss.* In a comment which suggests class solidarity between the editor and the Millers it was noted that *during all that time Messrs Miller kept open their works as far as they possibly could* and John May Miller actually used his own money to pay the men. Wages at this date were about £1.3.0 per week on average for 6 days at 14 hours a day though we are told *The work is not hard.* Boys earnt between 7/6 and 19/- a week and women from 7/- to 11/-. The Journal piece ended by recalling the previous lock-out in 1834 and noting that the factory was still working with the help of a few non-striking men and about 100 boys and women.

Hopes that the dispute would be quickly settled came to nothing as the following week the Journal reported that the men were still out.[18] The factory was still running but at a much reduced capacity. The strikers had apparently received no financial help from the Nottingham Union but *it is said that a delegate is expected therefrom with funds in a day or two.* The report again included a note of sadness about the mens' action reckoning they ought to consider *whether constant work, with wages far more liberal than they formerly had, is not better than the problematical advantages they anticipate from joining the Union.* This was the first mention of increased wages and indeed the only mention.

A week passed and still the strike remained solid.[19] The men had received some 16/- each, with boys getting 10/-, from a Mr.Axhorne of Chard which was a contribution from their fellow workers at that place and Nottingham. At a meeting in the White Horse Inn to distribute the money Axhorne alleged that the wages of the Barnstaple workers *were fully 35 per cent below what they ought to be.* He was thanked by *two spokesmen of the lace hands Mr.Beer and Mr.Cockram.*

The deadlock continued into October with the employers refusing to agree terms and the men receiving weekly financial support from the Nottingham and Chard Unions.[20] By this date, however, about 12 blacklegs had been taken on by the Millers and *between 40 and 50 machines – being fully one half the machinery in the mill* was at work. The Millers' only concern was to maintain the *high reputation in the trade* of their goods.

In a last ditch effort to smash the strike the Millers took the drastic step of taking 22 of the strikers to court on the extremely grave charge of conspiracy to halt their masters' business and also of *unlawfully, riotously, and routously disturbing the peace.*[21] The prosecuting lawyer was a Mr.Ffinch of Barnstaple and defending was a Mr.Heath, a solicitor employed by the Lace Makers Union in Nottingham. The case, which lasted 7 hours, was heard in Barnstaple Guildhall which was *filled to over-flowing.*

After some initial legal sparring between the two lawyers Mr. Ffinch called his first witness Henry Walrond, a clerk at the factory. He began by estimating the workforce of the factory as around 300. Soon after the strike began he had seen *rows* occurring between those on strike and those still at work. The former regularly called out *Ba-ah, Ba-ah, Black Sheep!* to the latter on leaving work. He then identified those he had seen.

The next witness was John May Miller who in a long examination was forced to admit that he had sacked the men just for joining the Union and knew of no misconduct they had been invloved in. More crucially to the conspiracy charge it was noted,

Witness could not answer the question as to what way the defendants had impoverished him or intended to impoverish him.

Bessie Dunsford then took the stand and alleged she had been insulted and pushed around by the strikers when she went into work at the factory. She was followed by a succession of others who all alleged intimidation from the locked out men.

Mr.Heath then addressed the magistrates at some length about the right of all workers to join a Union and withdraw their labour if they wanted. He accused John May Miller of *the intimidation of workmen by their masters* which caused a *slight sensation* in the packed court. As to the evidence about alleged intimidation by the strikers this was no more than *vulgar language on the part of a few boys, and hasty expressions in the nature of threats on the part of one or two of the men.*

After this the magistrates withdrew for half an hour and on their return announced that they had heard no evidence of conspiracy and thus dismissed all the defendants with *a caution as to their future conduct.* The result was met with *repeated and long continued cheering* it being noted that there *was no attempt to hiss the prosecutor, or any of his witnesses, as they left the court.*

This victory was followed by a letter to the Millers from the Union suggesting an interview which might lead to *some modification of your opinion respecting the Union.*[22] This olive branch, however, was peremptorily answered by the Millers who restated their *determination to employ no members of the Nottingham and Chard Society of Operative Lacemakers and from this decision we do not intend to recede.* Not surprisingly this intransigent attitude led to the out-migration of many of the Barnstaple men apparently to Chard if a contemporary report is to be believed.[23]

That some men persisted in their strike is shown by a report from December, some 2 months after the trial when the Union members met at the White Horse Inn to receive ten sovereigns from the Operative Stonemasons' Society of England, Ireland and Wales.[24] The chairman William Cockram added that to date the Derby laceworkers had received some £400 from the Nottingham union members and *if they wanted £4000 they could have it, so long as they stood firm to the Society, and that twenty other Trade Unions had offered to support the Barnstaple hands.*

These proud words, however, seem not to have been enough to keep the mens' resolve. Whether it was the Winter cold, hunger or boredom it is clear that the men had lost the will to continue the fight. At the end of January the Journal reported that following one man's return *many others followed after a few days*[25] though there were *some of them whom the employers would not again receive into their establishment.* Thus ended the second attempt to establish a union amongst the Derby workers. The whole sorry episode shows just what a powerful position Victorian employers occupied. The Millers do not seem to have been any worse (and in many ways they were better) than their contemporaries. That the Millers weren't too worried about the strike is perhaps shown from John May Miller's purchase of Rawleigh House in Pilton at the height of the action.[26] The old house was to be *his future residence* having long been in the Chichester family.

Within six months of the end of the strike 50 laceworkers were being taken on a *Wayzgoose* to Ilfracombe for the day.[27] Here they drank toasts to their foreman Mr.Trist and presented a silver cup to Mr. Garland *for his services as secretary during the last three years.* These trips became a regular feature of life at the factory and many are reported over succeeding years.[28]

Over the next two decades there is little recorded about the factory or the Miller family. In 1877 John May Miller was re-elected as a Guardian of the Poor.[29]

William remarried in 1880 as was noted earlier and in

1886 John May Miller left the Liberal Party to join the Unionists following a major national debate about the future of Ireland.[30]

In 1889 H. Strong published his collection of articles on the industries of North Devon.[31] He gives the number of workers in the factory at this date as 250 noting that this total included both *indoor and outdoor hands* they making *an important contribution to the commercial prosperity of the locality.* Some of the young employees were 'half timers' i.e. they went to school and worked in the factory in equal proportions.

The works *together with the accommodation land attached to the factory* covered over an acre with the main buildings forming three sides of a square. The right wing contained *new rolling locker machines* whilst the left provided premises for the *commodious counting house of the firm.* The main bulk of the building was the spinning mill. Within the factory all the workrooms were heated by hot-water pipes with ventilation provided via a *Blackman air propeller.* The buildings were claimed to be fireproof and were until the disastrous fire of 1972 did a great deal of damage.

In the main mill, according to Strong, Bengal and Italian silk was put onto bobbins before going to make the net the factory specialised in. This silk needed delicate handling unlike the more robust cotton which was brought from Lancashire via London courtesy of the South Western and North Western rail companies. The finished product from the factory at the time Strong wrote his report consisted of Chantilly and Cambray style lace in silk and Mosquito, Brussels and Mechlin in cotton. This was sold on to other factories where the finishing was done, the half completed lace being described as *in the brown.* Much apparently went to Switzerland.

A few years after Strong's article in October 1893 the Millers converted their firm into a Limited Liability Company.[32] Capital was £60,000 with the three brothers becoming directors and W.M.Jones the Manager and Secretary. No shares were offered for sale to the public which indicates the wealth of the Miller family at this date.

The company seems to have quietly prospered as around 1899 Alfred Miller announced his long-held intention of building an *Institute* at Derby to provide reading and recreational rooms as well as a Concert Hall and ornamental gardens for the factory workers.[33] His plans came to fruition and in October 1901 the red brick building was opened.

Press coverage was large including a whole page in the Journal which was headed *Institute & Pleasure Ground for Derby – Mr.Alfred Miller's Munificent Gift.*[34] Seen as directly following on from the *princely gifts* of W.F. Rock to Barnstaple the land had been purchased in 1899. Three cottages and the Newington Inn which stood on the site were demolished and the redbrick Institute constructed over the next two years following the laying of the foundation stone on 5th. of September 1900. The finished building provided a concert hall seating 300, reading and recreation rooms and a *refreshment bar* which sold beer though this was *strictly limited to the smallest quantity per member by the regulations.* In addition there were offices, a kitchen, toilet, a novel electric clock in an imposing tower and a caretaker's house, the whole building being centrally heated. The architect was James Crocker of Exeter and the builders Sanders and Karslake of Barnstaple. Total cost was £3077 plus another £2000 for fittings and land. The latter consisted of 4 acres of *Pleasure Ground* which included ornamental flower beds, bowling greens and a football pitch. All these amenities were intended *primarily for the benefit of the workpeople at Derby Lace Factory and their families* and membership dues were set at 1/- per year. The whole undertaking was to be run by eight trustees and the original trust deed said that the whole development would revert to the Town Council if the Miller company ever ceased trading in Barnstaple.

The opening ceremony began with Alfred Miller unlocking the main doors with a silver key and then making a speech of welcome. Most of his sentiments were normal for events like this but he did add,

I hope we shall never see a repetition of what has been witnessed in the Park so generously presented by the

late William Frederick Rock to the inhabitants of Barnstaple – I mean the defacing and injuring of seats by foolish persons cutting their initials on them.

After this proof that human nature never changes he then listed the many people who had helped bring his dream of an Institute to fruition and ended with the personal admission that his *infirmity of deafness will prevent my hearing anything* of the speakers that followed him. This meant he missed speeches from the Mayor, W.M. Jones and the local MP. The ceremony ended with the presentation of an album containing the signatures of the 400 or so employees thanking Miller for his generosity and the raising of *an imposing flag bearing the name of the Institute.*

Under the original Trust document Alfred Miller remained entitled to hold the office of Chairman of Trustees until his death or resignation.[35] The Trustees could appropriate any building on the site for use in *such a manner as they thought conducive to the innocent recreation of the.... employees and their families or any of them or to their moral and intellectual instruction or their social intercourse.* The Trustees could also permit *any person not being an employee of the said company or one of his or her family but residing either temporarily or permanently in Barnstaple.... to share the enjoyment or any of the privileges.*

The Primary Trust period was for *whichever shall be the shortest of the following two periods,*
1. *the lifetime of the longest liver of the lineal descendants now living of Her Majesty the Queen and a term from 21 years from the death of such longest liver OR*
2. *until the said company shall be wound up either compulsorily or voluntarily or without having been so wound up shall have ceased for 12 calendar months in succession to carry on their manufacturing at the Derby Lace Factory or at some other factory or workshop within 10 miles in a direct line from the Town Hall.*[36]

One piece of oral history suggests that the factory workers didn't respond to the high ideals of the

Institute's builder as he might have wished. A Mrs.Watts thought that Alfred's *efforts to get the men away from their betting and pigeon racing was in vain*[37] and quoted as evidence that the men used catapults to break the electric clock. A better response was reported from another employee who noted that, *Many workers went to the Institute in the evenings for further education as most had left school before the age of fourteen.*[38]

Whatever the truth of the matter tragedy struck the company during this period when, in June 1899, John May Miller died at his retirement home in Sidmouth.[39] By his death *the poor and various organisations* were said to *have lost a true friend and wise counsellor.* Over 100 people attended his funeral and amongst the wreaths were four from the Barnstaple lace workers.

Running of the company passed to the two surviving brothers but only two years later William died at his home in Nottingham.[40] Described as *a man of fine physique* he succumbed to bronchitis aged 72 and was, unusually for this date, cremated. His son R.W. Miller, a Nottingham solicitor, succeeded him as a director of the family firm.

Another two years later in May 1903 the last brother Alfred died.[41] Like William his body was cremated having to be taken to Manchester which seems to have been the nearest crematorium to Barnstaple at this date.

Derby Lace Factory c1920

THE MILLERS BOW OUT

In the first decade of the twentieth century the factory appears to have experienced a small boom. In 1906, for example, a great deal of overtime was worked at the Derby factory[1] whilst 1909 saw further development of the factory buildings.[2] The First World War saw the Barnstaple lace industry go through a series of changes. The first month of the war had the Mayor of Barnstaple remarking that,

> ...he very much regretted to hear that the Derby Lace Factory, which gave employment to many locals, was shortening the time of employment to only one day a week. This necessarily meant that there would be a lot of poverty, and they would have to make provision in some way to meet the distress which would be occasioned.[3]

Somewhat oddly, however, in the same issue that reported this there was news of a large American order.[4] The Journal's *Review of the Year* for 1914 noted that this order saw the factory working flat out.[5] This was repeated the next year when it was noted in December that there *has been for several months exceptional pressure at the Lace Factory.*[6]

The actual ownership of the company is hard to follow at this time but by 1915 William H. (or H.W.) Miller, another of William's sons, was described as *Director of*

Miller Bros.Ltd. in a report on the death of his son during the First World War.[7] This W.H. Miller was then living in Exmouth and it would appear that there was no male member of the Miller family then resident in North Devon. Everyday running of the factory seems to have been left in the hands of the Manager. As noted this was initially W.M. Jones but latterly W.H. Davie.

In terms of trade after the boom years of the First World War it was perhaps inevitable that business would decline but the closure of the factory in November 1920 seems to have been unexpected. The Journal coverage is small referring as it does to a *grave announcement* from the management.[8] Apparently owing to the *extreme depression in trade, and the accumulation of stock* all piece workers were laid off as from Monday 1st. November whilst all other workers were given a week's notice. Barnstaple wasn't alone in experiencing this depression as many other similar factories were closed or closing. One item of interest in this terse report is that *all the lace making machinery has been made by the firm's own staff of engineers and fitters.*

The closure led, as might be expected, to widespread hardship in Barnstaple among the workers and their families. This became so pronounced that a Distress Fund was set up by the Mayor following a public meeting in the Town Hall.[9] At this a report compiled by a committee of council officials plus *two employees at the Derby Lace Factory* was read which highlighted the problems being experienced by the unemployed. Street collections by the workers had raised about £75, including £10 from Reginald Miller one of the factory's directors. This money had been spent on relieving the worst cases but still the children at St. Mary Magdalene School in Derby were *half-starved* according to the headmaster.

The meeting appointed various people to run the Distress Fund and restart the Barnstaple Soup Kitchen. Additionally local people in work were to be asked to copy the example of the men in one Barnstaple factory who had volunteered to give 6d a week each to help their unemployed brethren.

Within a short time the Fund stood at £545 and

weekly lists of those who had contributed were being printed in the Journal.[10] The Fund eventually raised some £1300[11] but this tremendous effort came to an end in June 1921 when the Journal reported with *great pleasure* that the Derby Lace Factory was being reopened *as a few orders are to hand.*[12] The restart must have been slow as only some of the machines were being brought back into operation but it was *hoped that some additional orders will be received and that ere long all the machines will once more be put in motion.*

The business seems to have continued but not at such a rate as before and with the world depression in the 1920s it once more went through hard times. In April 1925, for example, the factory switched to short working with the average day being cut from 8 hours to 5.[13] When this was announced about one hundred lace twisters and threaders stayed off work in protest. The Journal noted that this shortening of hours was *necessitated purely by reason of immediate conditions.* The manager W.H. Davie responded by posting a notice at the factory gates indicating that absentees were *liable to prosecution as well as instant dismissal* and that if the men didn't return to work the other 300 employees would have to be laid off.

The men stayed out for a week and then returned after discussions between H. Gammidge *organiser of the Workers Union* and the management when it was agreed to reinstate the 8 hour day.[14] Almost as an aside the report in the Journal noted, *The firm agreed to recognise the Workers' Union, and requested that body to elect delegates to meet the management should any further difficulties arise.* Clearly the world had moved on considerably since the unsuccessful strike over union recognition some fifty years earlier.

In June 1929 the factory was again closed owing to a shortage of orders. The Journal in reporting this noted that the factory had relied for many years on large foreign contracts but *certain of these countries have resorted to manufacturing their own lace.*[15] In addition to this *competition in the industry has increased* and one recent order for mosquito netting which the Barnstaple factory had expected to secure was actually placed with a Nottingham firm. The closure, however, was only expected to be temporary with the 3-400 employees soon being expected to be back at work on the *hundreds of the most up-to-date machines for the making of lace* which the factory housed.

Two months later, however, the factory was still closed and the only good news the Journal could print referred to a *plain net mending* business being established in Summerland Street in the town by Sir Arthur Black, head of a Nottingham lace firm.[16] Unfortunately this didn't employ many people but at least it provided some income.

The seriousness of the situation was brought home to North Devon when in August 1929, two months after its *temporary* closure the Derby factory was put on the market.[17] Commenting on this the Journal said it was *the result of contemplated readjustments in connection with the Company, which has its headquarters at Nottingham.* Reference was also made to *the stoppages in the cotton industry* which had meant a curtailment of supplies of raw material. It would appear that the Miller family were, after a century, returning to their home area. Apparently the Millers did in fact cease trading altogether in the 1920s as they do not appear in the Nottingham directory for 1928.[18]

That the Barnstaple factory was still an attractive proposition even during this problematic period is shown by the fact that a purchaser was found within a few weeks. Messrs. Small and Tidmas were the buyers, they already owning lace works in Nottingham and Chard.[19] An interview with J.C. Small elicited the comment,

We have bought out the factory, and are naturally giving matters very sympathetic consideration with a view to continuing its running; but to some extent our decision will be governed by the way in which our efforts are construed by other bodies.

This rather cryptic comment was thought to be a veiled hint to the Town Council that the rates on the buildings making up the factory needed to be lowered if the business was to become viable again.

These problems must have been overcome as a Mr.F. Dobbs, the resident manager, announced the re-opening of the factory on the 21st. of October 1929 when some 60% of the workforce were to be immediately re-employed with a hope that the remainder would be taken on as soon as possible thereafter.[20] In its coverage the Journal ended by saying,

> *The news has, indeed, afforded unbounded delight in the Derby district, where a large proportion of the employees reside; and we are informed that the re-opening of the Mills on Monday morning is eagerly awaited.*

After the Derby Lace Factory was sold in 1929 the Miller Institute seems to have been used as a school on a temporary basis at least until 1933 when Hubert Miller began protracted negotiations with the Town Council as to its future. The Council, however, were unwilling to take over the building on a permanent basis. Hubert Miller regretted that the Council could not *secure the premises for educational purposes* and stressed throughout the suitability of the building for education, especially of the young unemployed.[21] Eventually of course the situation was resolved and the Miller Institute has continued, as its founder wished, in being a place of education by housing, in turn, the Barnstaple Boys' Secondary School and now Yeo Valley Primary School.

5

*Bobbin and lace
machinery*

OCCUPATIONS

As no wage books or employment records survive
from the factory itself, it remains to search
through the Census returns to gather an idea of
the number of lace factory workers who lived in the Derby
area, and to gain a picture of changes over the fifty years
which they cover.[1]

This, however, is not as straightforward as might first
appear, due to a number of factors. The variety of trades
practised at the Derby Works, which was completely
self-sufficient and not only manufactured the bobbin
net, but was also responsible for the innovation and
fabrication of its machinery, included workers like
smiths and carpenters as well as clerical staff such as
accountants. These are not always obviously identifiable
in the Census as being directly linked with the factory.
Strong's vivid account of the operation of the Derby
Works merits reading in its own right for his enthusiasm
alone.[2] In terms of occupations he firstly describes the
weighing and testing of the silk and cotton thread for size
and strength. The silk thread was then wound onto
bobbins by machine girls *as nimble with their fingers as
they are alert with their eyes* before being re-weighed and
tested and checked against their original batch ticket.
The cotton thread was damped to make it easier to work.
Indeed in the early days of cotton Manchester's famously
damp climate was a major factor in that area's suitability

46

for the development of the industry, as well as its proximity to the port of Liverpool and its trade with the Americas. Perhaps Barnstaple too, further down the west coast of England, and with its own established trading links, was seen to share this climatic propensity which might have encouraged the introduction and success of a cotton-based industry. However, by Strong's time in the 1880s, a damping department and steam powered centrifugal drying machine aided this part of the process.

After the second weighing, part of the thread went to the warping room. The lace warper would set his machine according to the length of warp required so that appropriate lengths of thread, or lays, were transferred to the rollers used in the lace machines. The woof, or weft thread, was carried in a brass wheel bobbin which was wound by automatic machinery. The Census shows many instances of wheel winders and brass bobbin winders, often very young children, especially girls, and it is probable that this was a job done by hand for at least the early part of the factory's history.

Another occupation recorded for the very young, in this case, young males, was that of threading boy. Strong describes the adeptness of the lads who *take the end of the bobbin thread and simultaneously carry it through the eye of the carriage in which it works, catch the spring, and place the bobbin within its circle as the delicate spring comes back to hold it in its place* – as many as 3900 carriages having to be filled for each new piece of net, of which three could be completed within two days.

Silk differed in that after the 'dramming' (weighing and sorting) and being wound onto wood bobbins, the threads were drawn over a drumming frame, a black board, which process would show up any imperfections, before being wound onto drums from which the warp would be made and the bobbins for the lace machines wound.

Strong calls the operatives of the actual lace machines *lace hands* or *lace makers*, both terms being found in the Census, but the most common occupation recorded is that of *lace twister*. This may be taken as a reference to the process of twisting by which the lace was actually made, the thread being laid down from the bobbins on one side and then the other of the warp threads, rather than under and over them as in weaving.

The lace net was then taken to the wareroom to be weighed and registered, mended if necessary, and finished. Silk menders were required to work on the premises, whereas cotton nets were mended by outworkers. Lace menders were nearly always female, and could be either married women or single girls living at home with their parents. Once checked by the overlooker, the nets would be folded to be packed for despatch. This would be done by female hands during the day, and by the watchman at night.

In Strong's time the goods were sent by rail, and *by special arrangement with the railway authorities, the lace goods despatched in the afternoon at three o'clock are delivered at the wholesale house of the firm in Nottingham at seven o'clock next morning.* No mean achievement by today's standards, and even more impressive that this was via Waterloo and Euston, not through Birmingham as would be the route today, and involved two different railway companies.

The factory was powered by steam, by way of two 50 horse power boilers driving a 60 horse power condensing engine. The blacksmiths' and fitters' shops and the carpenters' and pattern makers' shops were where both the maintenance of existing, and the construction of new machinery took place. *Wood models of every description of machinery in work on the premises are stored here* states Strong. Thus it is more than likely that many of the carpenters, smiths, and wheelwrights recorded in the Derby area were also employed at the factory, although this is not made clear by the Census itself. Other staff employed at the factory included those in the counting house; the accountant, clerk, factory manager, and in later years, the company secretary.

Reading through the 1891 Census – that most contemporary to Strong's account – we can put names to many of the occupations which he described. Princes Street, one side of which was largely owned by Miller Brothers at this time, was home to a full range of factory workers. Many of the women, particularly wives and daughters, worked as cotton winders. Bessie and Annie Biddle, whose father Robert and brother John were lace twisters, Annie Boaden, whose husband John and four sons were also lace twisters, to name just a few. Elizabeth Pile, aged 65, and Florence Perryman, aged 18, at opposite ends of their working lives, were both lace menders, and probably worked at home. Other women performed what would be regarded as more skilled work. Bessie Handford and Elizabeth Woollacott were both lace twisters and heads of their own households, perhaps indicating a certain independence reflecting their status. Elizabeth Heathfield, a long time Derby resident, worked as a silk drummer.

Elsewhere in Derby are recorded numerous lace twisters and menders, and others such as Maria and Martha Bubyer, mother and daughter, of Union Street, one a silk winder, the other a silk drummer. In addition to those occupations already mentioned there are Richard Pearce, machine fitter; Robert Jordan, blacksmith; and many loosely described as mill hands and machinists.

Job titles in the earlier Census returns are more descriptive. Others from the 1881 Census include John Fear, engineer in lace factory; Thomas Seager, timekeeper; Harriet Eames, wheelwinder; Ann Passmore, female overlooker lace factory; Philip Gardiner, overlooker in lace factory and Thomas Sutton, watchman lace factory.

The more precise terminology, particular to parts of the lacemaking process, may also reflect that at this time these tasks were performed by hand. Their absence from the later Census returns and the more general terms used may be explained by the development of machinery which made them obsolete.

In addition to Strong's account, contemporary newspaper reports also give details of working practices at the factory. These suggest that, for at least a part of the factory's history, the lace twisters were themselves responsible for the employment of the boys who threaded their machines, and also of the outworkers who mended the lace net they produced. One such instance has already been cited in Chapter 2 where John Miller quoted the wages paid to John Colwell, one of his lace twisters, who had to *pay a boy* to assist at his machine.

As far as the lace menders are concerned two Journal articles indicate changes in the way they were employed. James Moore, a lace twister was charged in November 1839 with non-payment of wages to Harriett Potter, a lace mender.[3] She stated that *she had been employed by Moore in the month of August last in mending lace for fourteen days and a half at eightpence a day, and that he now disputed the debt and refused payment –* a total of nine shillings and eightpence. Moore's wife, speaking for him perhaps due to his absence at work, or because she managed the money, said that she was *willing to pay Potter for thirteen days and a half which was all she owed her. The Bench remarked that the complainant had sworn to the amount and ordered Moore to pay the sum demanded in instalments of two shillings and sixpence per week.*

However, by 1842 the employment of the lace mending outworkers seems to have become more formalised and to have been brought under the control of the factory management rather than the lace twisters themselves.

In February of the following year 17 year old Thomas Reeves was charged with *having embezzled a considerable sum of money from the Derby Lace Factory.*[4] The Journal article goes on to explain in detail the procedure by which the lace menders were employed,

Mr May, the confidential agent of the trustees of the late Mr Miller [Miller having died in the previous October] *stated that the prisoner had been employed in the warehouse, with a woman, to receive the pieces of lace when brought home* [sic] *from the menders, to enter them, and make out a check for the amount of wages for mending, which the cashier in the counting house paid; these checks were afterwards compared by the cashier with the piece book. The persons who mended the pieces generally lived in the neighbouring parishes, and sometimes, in the absence of the cashier, they would, in order to return home, get the checks, which varied in amount from 2s – 10s, paid by some one on the premises who would afterwards receive the amount from the cashier.*

As may be guessed, young Reeves had seemingly taken advantage of this arrangement to *present, and get others to present fictitious checks, and by this means had embezzled during 12 months sums amounting to upwards of £68.* Suspicion had been *excited against the prisoner* at Christmas *from reports of his extravagant expenditure, and having absented himself from the factory several days without leave, he was discharged.* Subsequent checking of the books had *left no doubt of his delinquency.*

Reeves was sent for trial at the next Quarter Sessions in April at which he pleaded not guilty to the single charge brought against him *from a great variety of similar cases* of obtaining five shillings and sixpence on false pretences and with intent to defraud the trustees of the factory.[5]

At this hearing the lace mending system and its accounting was explained more fully,

> There was kept a book in which every piece of lace was numbered and entered as it came up from the loom; a menders' book on which a Dr (debit) and Cr (credit) account was opened with every woman employed in mending, and a sort of index or check book in which it was the prisoner's duty to copy from the menders' book the names of the women employed and the amount done on each piece of work sent out. When the pieces were returned, after they were examined and approved, the prisoner had to refer to the check-book and make out a ticket, bearing the name of the mender, the money due, and certified by his initials, which ticket was taken to the cashier, who paid the amount.

It was claimed by Sarah Cure, a lace mender and wife of Robert Cure, labourer of Derby, that on the occasion in question, in November 1842, Reeves had asked her to take a ticket to the cashier, James Gifford, on his behalf. He had stated that he had paid out another woman, *a country mender of Swymbridge,* called Dart earlier in the day but *had had some angry words with Mr Gifford and did not like to take it in.* She had then taken it with her own ticket and had given Reeves, who was waiting by the wareroom door, the money, – *it was two half-crowns and a sixpence.* In cross-examination Sarah admitted,

I can read some writing but not all kinds; I can't write; the time I received the ticket was half-past four or later; it was not dark, but quite light enough to see my writing; I could read the name of Dart; I thought it quite a [sic] honest transaction at the time, but have since had reason to think otherwise; I live at Derby; the cashier sent for me some time after the transaction – perhaps two or three months afterwards; he made enquiries of me, and I at once told him of the ticket I had taken in to him for prisoner; I told him the name was Dart and that if his file were right, he would find the tickets of Dart and Cure one following the other; I am sure that I took the ticket for Thomas Reeves in the month of November, for I remembered that it was in the month in which my birthday occurred; it was after my birthday which is the 4th of the month, but can't say if it was the 22nd.

This statement shows that then, as now, a lack of formal education does not indicate a consequent lack of street wisdom, especially where money is concerned. It also indicates something of the relative lack of awareness of time of day or year compared to our present-day society. It was the dawn of the railway which brought the country together on one time zone, prior to which there were many "local times". The rising and setting of the sun and the factory bell would have been the signs by which the people of Derby regulated their day.

James Gifford added in his own testimony that it was the duty of a woman, Martha Ward,

...to examine the lace to see if it be properly done, after which the prisoner was employed to affix the date to the piece on the menders' book, which was an acknowledgement of its accuracy, the menders give notice when they will want their tickets, and they are prepared for them, and they bring them to me. I pay the amount, and retain the tickets as my vouchers.

The statement given to the Court by Arthur Packer

Stevens, who had at this time worked at the factory for 12 years, indicates that the work of an accountant was not solely concerned with figure work as we think of today. He stated that *we have 150 menders employed, or more; my duty is to receive the pieces from the men, and to fix a price for mending upon them: the prisoner gave them out to the menders.*

Surprisingly, given the above, the defence counsel, Mr Bencraft, *objected to the loose manner in which the books of the factory were kept.* Perhaps he meant that despite the apparent organisation of the system, in practice there must have been a delay in, or lack of checking the books against each other to give Reeves the opportunity to appropriate such sums, and over such a period of time. He *besought the jury to give the prisoner the benefit of the doubts which they must necessarily entertain of his guilt.*

The jury retired for five minutes, and then returned a guilty verdict on the second count but recommended *mercy on account of his extreme youth and because the lax manner in which the accounts were kept gave him facilities to commit the offence!*

Equally amazingly the prosecutors instructed their solicitor, Mr Chanter, to *join the recommendation of the jury, because the father of the prisoner had been in the employ of Mr Miller for a long series of years, and had borne a blameless character.* In passing sentence the Recorder,

> ...remarked on the ingenuity the prisoner had displayed in his offence, which he had opportunities of committing only because of the implicit confidence which was reposed in him, and the abuse of that confidence was a material aggravation of his guilt. But for the recommendation of the jury, mercifully concurred in by the prosecutors, he should most certainly have sent him out of this country; but in guarding him from transportation he could not forget that his offence was a serious one, and it was due to the public safety that it be punished severely.

Thomas Reeves received 12 months imprisonment to hard labour.

As Thomas was saved by his father's own good reputation as a worker at the factory, many other families in the Derby area had members of more than one generation employed there. Indeed, at various times in the fifty years covered by the Census, examples where entire family economies were dependent upon the factory as a source of income may be found.

The Fisher family of Higher Maudlin Street consisted in 1861 of,

William	Head	45	Lace twister
Maria	Wife	45	—
Samuel	Son	20	Smith
Thomas	Son	18	Lace twister
Susan	Daughter	16	Cotton winder
Selina	Daughter	14	Cotton winder
Albert	Son	12	Threading boy
Frederick	Son	10	Threading boy
Ellen	Daughter	7	—
Isaac	Son	6	—

We can only speculate as to whether Samuel worked in the factory smiths' shop, or whether Albert and Frederick were the boys responsible for threading up their father and elder brother's machines. As such they may have been saving the family income rather than adding their own earnings to it, avoiding the need for their father and brother to pay other boys to do their threading.

Similarly the Hooper family of Union Street recorded in 1861,

William	Head	35	Lace twister
Fanny	Wife	35	Lace warper or worker
Ann	Daughter	14	Cotton winder
Esau	Son	12	Cotton threader
Jacob	Son	12	Cotton threader
Eliza	Daughter	9	Cotton winder
William	Son	5	Scholar
Robert	Son	6m	

There are also instances where the factory supplemented the main income. Thomas Garland, a mason employing four men and one boy in 1851, who had himself been involved in the building of the Derby streets, lived in Union Street with his wife and seven children and a nephew. His four daughters, aged 21, 16, 14, and 11 years, are all recorded as lace menders. This was something they could all do at home and perhaps maintain the 'respectability' associated with not having to go out to work, but still bringing a welcome income into the home to help support the large family.

CHILD LABOUR
The 1841 Census does not attribute occupations to many children – the youngest lace worker living in Princes Street was William Woodward aged 12, whilst of Boden's Row John Colwill aged 11 was listed as a lace twister.

By 1851, all children apart from infants were listed either as a scholar or given an occupation. William Richards aged 10 and his sister Sarah, 12, were both described as cotton bobbin winders, their elder brother John, 16, and father, also William, were both employed as lace makers. In Boden's Row, William Cockram, aged 7, was employed as a lace threader whilst his elder sister Emma, 9, was a scholar. Another sister, Hannah, 11, was a lace worker, but neither parent was connected with the factory, their father being an agricultural labourer. There seems to be no obvious logical reason why parents appeared to send some children to work, whilst others stayed at school, often well past the age that their siblings started work. A possible explanation is that the term 'scholar' disguised the activities of children who were actually working in the factory at least on a part-time basis, and that the use of child labour was much more prevalent than the Census suggests. It could also be that parents did not necessarily want to disclose their children's activities to a government official.

There is mention in various Journal articles of January 1834 to a school existing at that time in Derby, on the same lines as which it was proposed to start another in the town. These discussions, over three issues of the newspaper, stated that;

> ...it is desirable to establish an infant school, on the most unsectarian and liberal principles, and that the committee of the school at Derby be invited to co-operate therewith [6] and that the proposers were heartened from the success of a similar institution on a limited scale at Derby.[7]

The liberal non-conformist John Miller is quoted as having no objection to the mistress of the new school as being a member of the established church, rather he would vote for it, as the mistress of the infant school already subsisting was a dissenter.[8] Miller's involvement in this debate could have been solely due to his position on the corporation, but it is also possible that the school could have been associated with the factory, particularly as its mistress was, like him, a non-conformist. Certainly Mrs. Miller as already noted gave a treat to 75 children belonging to a charity school at Derby in 1838.[9] If such a link could be proved it would challenge the claim of that of Heathcoat's factory at Tiverton that its factory school, opened in 1843, was the first in the West Country.

The scholars of Derby however were not as appreciative of their education as the town elders. An 1838 Journal report records how,

> Miss Caroline Hedger, mistress of the infant school at Derby complained of 5 boys viz – James Nutt, Henry Petters, John Pugsley, William Hill and Joshua Diamond (most of them belonging to the Derby and Rawleigh Factories) for molesting and assaulting her.

Whilst she was walking through Derby they had,

> ...assailed her with stones and dirt, which they threw into her face, and about other parts of her person, accompanying the assault with the most improper language, and two of them were pushed up against her by their companions.

Miss Hedger could only identify Petters and Pugsley, presumably through association with the school, and they were fined 9d each with 3s 6d expenses – the others being discharged with a caution.[10]

The reluctance of the part-time students is echoed in the account of a meeting of Barnstaple School Board in 1873, when *Mr Miller remarked on the great difficulty in getting the half-time children in the factory to attend regularly. They would come to work but they hated school.*[11] An officer was to be sent to the factory to *take the names of the boys there who ought to be at school, and see that their attendance was enforced if necessary.* These reports only refer to boys – whether girls also attended school, or whether it was just the boys who were troublesome is not clear.

The Rev. Boggis refers to St.Mary Magdalene School, opened in Lower Maudlin Street by his predecessor the Rev.Bull in May 1861. This was to be *for ever hereafter appropriated and used as and for a school for the education of children and adults or children only of the labouring, manufacturing, and other poorer classes in the parish of Barnstaple.*[12] Separate accommodation in Union Street was later used for the infants until 1865 when this was added to the main school site. He also mentions the Wesleyan Day Schools with Mixed and Infants departments in Reform Street as having been opened in 1853 and schoolrooms associated with the Methodist church in Bear Street and with the Brethren Chapel in Grosvenor Street.

He recounts that,

...the teachers' efforts are now no longer trammelled, as used to be the case, by that hardest of hard tasks – the schooling of half timers, big lads who worked in the lace factory in the morning, and were obliged to attend school in the afternoons, and who were so strong and unmanageable that one head master is said to have had a leg broken in a struggle to overcome one of them.[13]

Early concern about the effects of factory life on the habits of children had been voiced in a Journal report of February 1829. This involved one John Symons, aged 11 years, and another boy named May of about the same age, who had been sent to jail to be tried at the next

Quarter Sessions. Their heinous crime was the theft of three quarters of a pound of soap! This rather harsh treatment was qualified by the statement that,

this case presented a complexion rather novel here, but of frequent occurrence in the metropolis, and other places of large population, and is perhaps the result of the association of youths employed in manufactories.[14]

The Quarter Sessions report, from May that year, revealed that the involvement of a third boy, *an accomplice, who stood outside the door, proved that it was a preconcerted scheme between them to go to this shop and steal the soap.* This evidence of premeditation led the boys to be warned that *if they were again found in the situation in which they then were, they would be transported.*[15] Instead Symons was sentenced to a further month in jail, and May three weeks, both with hard labour. The bench appeared determined to make an example of them and so hoped to put an end to crimes involving such groups of working boys.

However, this was evidently not to be. A later Journal article reporting the stoning and theft of a duck by three factory boys from Mr Ching of Frankmarsh Farm, quotes the owner as stating that,

...this was one of a very great number of depredations which the boys of the factory were accustomed to commit on the Sabbath day, when they usually assembled, sometimes 100 of them together, and sallied forth to the neighbouring orchards and gardens to the great annoyance of the proprietors.[16]

Certainly the lace boys appear regularly over the years in Journal reports of court cases involving theft of various foodstuffs from nearby growers – mainly apples, from Pitt orchards[17] and Westaway[18] for example, and more unusually, walnuts from a garden at Rawleigh.[19] For this last crime James Bennett *who was employed at one of the adjacent lace factories was committed to the treadmill for a month.* Given that he *had previously been found guilty of a similar offence when they defrayed the expense* his parents refused to pay a fine, hoping that *the punishment of a prison might operate on him somewhat beneficially.*

Two cases which were perhaps the result of boredom and lack of play space rather than real malice, were the bringing to trial of eight boys employed at Boden's lace factory for damaging a lilac shrub at Dr. Morgan's house and of five boys, in February 1835, for trespassing on Mr Snow's land.[20]

In the first case the boys were accused of *wilfully and maliciously damaging some shrubs in front of Dr Morgan's house.... It appeared that the complainant's property had been frequently injured by those and other mischievous boys.*[21]

In the second case, Mr.Snow stated,

...that for several years he had been subjected to a great deal of injury and annoyance, from the cause of which he now complained, but he had never troubled the court upon the business; it was impossible, however, that he could put up with it any longer; it was not uncommon for 50 or 60 boys, (on a Sunday especially) to collect themselves in his field to play, and so daring had they become, that if he sought to drive them away, they loaded him with abuse, and absolutely refused to leave, and in some instances had assailed him with stones and other dangerous missiles. He had no wish to press for any penalties, for he was sure their parents could ill afford to pay them; his object was to convince them that they had no right to trespass upon his property, and he hoped that the admonishments of the bench would check the evil.

The boys were fined one shilling expenses only, in lieu of the usual five shillings, after *expressing their sorrow for the offence and promising never to repeat it.*[22]

Mr.Snow lived at Vicarage Lawn, a large house in its own grounds, which stood where are now the streets of the same name. He ran it as a private boarding school, and one wonders if there was a certain animosity between the boys who lived and studied there, and their contemporaries who had worked long hours at the factory, often from a very young age. The latter must surely have envied the former their free time and space for play.

That the boys were often reluctant to attend to their work is confirmed by a report involving an alleged assault on a boy by an overseer at the factory. At the hearing John Miller himself *corroborated the evidence of the overseer, as regarded the idle and disorderly conduct of the boys, and added that he had been compelled to engage a man for no other purpose that to go through the rooms to keep the boys at their work.*[23]

The summing up of this case sheds an interesting light on the then conditions of the contract between master and apprentice, and of industrial relations in those days. In proving the assault on the boy, William Bevan, against the overseer, Nicholas Thomas, the court reasoned that,

...had the boy been an apprentice the master would have been justified in administering moderate chastisement but as he was not, they could not sanction the application of condign punishment in the event of improper conduct on the part of the boys, the remedy of the master was to bring them before the magistrates, who would administer suitable correction.[24]

Tensions between the younger members of the workforce and their elders are also reflected in other newspaper stories. In August 1834 the Journal carried a report of an alleged assault involving a ten year old girl lace worker.[25] An account from August 1840 describes an assault case brought by a lace boy, John Thorne, against John Glover, an overseer at the factory for *kicking and beating him with a rope*, the boy's mother bearing witness to the marks on her son's back as a result.[26] However, it appeared from the defendant's statement, *corroborated by the evidence of Mr Stevens* (the company accountant) and Anne Burgess who witnessed the affair that young James was not entirely innocent in the matter. He had *pushed down a girl named Mary Peters who had a quantity of thread in her hand, upon the stone steps, and made her mouth bleed.* She had then complained to Mr.Glover who had *struck the complainant but did not kick him.* James had then *made use of some of the most disgusting expressions* and Glover had then *hit him four times across the neck with a small rope.*

The magistrates dismissed the case as although *they would not countenance any undue severity... there did not appear at this instance to have been any more than whole-some discipline.* They concluded that *there seemed reason to suspect that the complaint had originated in a revenge-ful feeling on the part of the mother because the boy had been discharged from the factory for his misconduct.* Perhaps this was compounded by the importance of the contribution of child labour to the family economy.

The court's sympathy did not always lie with the adults accused in such cases. An earlier case, from December 1833, merits quoting at length to illustrate this. It reads,

> *Henry Burgess, an operative in Mr Miller's lace factory, was charged with a most barbarous and brutal assault on a poor little boy named John Featherstone who is also employed in the same factory. The complainant, a child about 12 years of age, stated that on Friday evening when he was at work at his machine, Burgess came and took from it a bobbin, that he went to him and asked him to return it, but he refused to do so, he then applied to the foreman, whereupon Burgess caught him by the throat and with a strap, beat him in a most unmerciful manner.*
>
> *Edward Baker, another lad who works in the factory, deposed that Burgess continued beating the com-plainant for 5 minutes at least and hearing him crying very much, he went to him, and found him lying on the floor in a kind of fit, when he took him up and assisted to carry him home to his mother.*
>
> *William Gliddon also bore a similar testimony with the addition that he saw Burgess strike him twice after he was on the floor.*
>
> *The mother of the child stripped off his clothes and his body exhibited such marks of severe punishment as excited feelings of sympathy for the poor boy and indig-nation against his heartless and unmanly punisher.*
>
> *The defendant alleged that the boy threw some irons at him but this assertion was disproved.*
>
> *The court decided that a most cruel and flagrant assault had been proved against the defendant and*

considered it a disgrace to the whole factory that a number of men, who were at work in the same room, could suffer a child to be beaten in such a shameful and barbarous manner without interfering to prevent it, and convicted the defendant in the penalty of four pounds and expenses, in default of which he be imprisoned two months.[27]

Interestingly in a gesture of sympathy towards the accused *the mother of the boy interceded for mitigation of the penalty but the court did not think it proper to alter the sentence.* Whether this reflected a common understanding of the financial implications or doubt as to the total innocence of her son in the matter cannot be known. One wonders whether the mother of John Thorne in the previous account remembered this incident which had occurred seven years beforehand. Ann Burgess, a witness in that case, was possibly a relative of Henry Burgess, the defendant in the latter.

That young children were engaged in employment at the factory is further confirmed by the newspaper reports of often quite tragic accidents which befell them at times. A report from November 1852 recalls an accident whereby *on Monday last, a little boy named Martin, about nine years of age, while employed at Derby Lace Factory, had one of his toes cut off by the machinery. The poor little fellow was conveyed to the North Devon Infirmary where he was promptly attended to, and is doing well.* Whether he was wearing anything on his feet at the time is not stated. The tragic chain of events suffered by this particular family is revealed by the additional comment that *He is brother to the boy who died a week or two since from injuries in the abdomen occasioned by running against the handle of a sweeping brush, and some years ago, another brother lost his life by drowning.*[28] The accident with the brush had not happened while the boy had been at work, but while chasing other children in the street but perhaps the other brother was a victim of the notorious plank bridge between Rawleigh and Derby which claimed so many young lives.[29]

Accidents involving other older boys include such as that of 12 year old John Verney who *whilst attending a machine, entangled his hand in the machinery, which drew it round the shaft, whereby his arm was fractured a little below the wrist. Fortunately by the breaking of a cord his hand was liberated without occasioning further injury.*[30] In another instance Thomas Light, 13, *whilst working at Mr Symons' factory.... got his leg entangled in a pulley and broke it in two places.*[31] In both cases these *little sufferers* were expected to recover. One marvels at their luck though, in those days before Health and Safety Acts regulated the workplace, when both medicine and machinery were crude, and when disability meant, in effect, an extra mouth to feed in the family economy.

New Buildings

HOUSING

As already stated the Derby district got its name from the area from which John Boden, the original founder of the lace factory, is thought to have originated. The part around the factory itself was originally known as Stoneybridge, and this remained its address for many years. The adjoining area on the factory side of Vicarage Street was the Brickfield or Prior's Bear, whilst that on the other side was Vicarage Meadow or Gaydon's Ground. Gaydon Street was once known as Short's Lane, perhaps being renamed when the houses were built in the 1840s? Parish records from 1828 refer to 'the Derby' and 'Little Derby' and eventually the name Derby became that of the whole area.[1] There is a saying in Barnstaple that Derby was where the policemen went in pairs, but whether this dates from the last century or is the result of more modern events is unclear. Certainly it is evident from contemporary Journal articles that Derby was a colourful place to live.

A popular belief is that the area consisted of 'workers' housing', with the implication that the houses were built and owned by the factory itself. In fact, the only street of which this can be said to be true was the narrow street mentioned earlier as having been built by John Boden known in earlier times as Pugsley's or Boden's Row, and later renamed Corser Street, apparently after his daughter who married the Rev. John Corser who at one time

64

1. The Raleigh factory c. 1880 showing the row of workers' cottages in the foreground that still exist today. The factory itself was almost completely destroyed by fire in 1888

(North Devon Athenaeum)

2. Derby Lace Factory c.1884 showing the original four storeys and the extension wing erected in 1874.

(North Devon Athenaeum)

3. Gorwell House in 1997. This imposing building was constructed by John Miller for himself and his large family probably in the 1830s.

(Deborah Gahan)

4. Barnstaple Guildhall at the end of the nineteenth century. It was here that the Millers attended council meetings and where the meeting to resolve the strike of 1833 and the 'conspiracy' trial of 1874 took place.

5. A studio photograph of John May Miller (1829-99) who served as Mayor of Barnstaple 1863-64 and ran the Derby Lace Factory from 1863 until his death.

(North Devon Athenaeum)

6. An interior view showing one of the lace making machines built by the Derby factory staff c.1884

(North Devon Athenaeum)

7 a + b. Two photographs from c.1884 showing the Bobbin Room and the Finishing Room in the Derby factory with female staff in their long white 'working' dresses. Note the many windows necessary to supplement artificial light when dealing with such fine fabrics.

8. a, b, c + d Four views of the lace making machinery taken in the 1920s showing the massive scale of the works at this date.

(HJ Pinn)

9. The Mill Staff 1883

(S&T. Barnstaple)

10. The Wareroom Staff 1883

11. Looking down Union Street c.1945-46

(North Devon Record Office)

12. Corser Street c.1945-46

13. Vicarage Street with entrances to Reform Street, Union Street and Newington Street. late 1950s

(North Devon Athenaeum)

14. Corser Street with its remodelled entrance, late 1950s

15. Looking down Newington Street towards the Union Inn with the factory chimney in the background, late 1950s
(North Devon Athenaeum)

16. An early aerial view of the Derby Lace Factory and surrounding streets

lived at Zephyr Cottage, just up the hill from the factory.[2] Rate books for the 1830s and 40s and from 1879 and 1910 confirm that this one street remained entirely in the ownership of one person or trust during these years. Susan Pugsley is named as their owner in an undated rate book from the 1830s which explains the origin of the alternative early name of the street.[3] In 1840 the owner is named as Susan Corser.[4] Presumably if Boggis is correct then the former Miss Susan Boden must have been widowed and remarried. She is also listed as the owner of houses in Green Lane, the occupants of which included John Tallyn and Thomas China, two men associated with the factory and who later lived in Boden's Row. Thomas China, a cordwainer originally from Tiverton, had moved there by 1851 when his sons Charles, aged 21, and Thomas, 19, were both lace twisters.[5] John Tallyn, himself a lace twister, was living at 15 Boden's Row in 1871, previous to this he is listed in Union Street in 1861. John Boden evidently did not include the housing when he sold his factory to John Miller although the street continued to house a large proportion of lace workers for the rest of the century. The 1879 rate book shows Boden's Row as consisting of 46 houses belonging to the Gammon's Trust, and a warehouse and garden owned by John Corser.[6] In 1910 Cobbs' Trust are stated as owning the houses.[7]

Boden's Row, shown on an 1829 map of Barnstaple as New Street, was then one of the first Derby streets to be built and differed from its followers in its style and construction. The original entrance was a passageway beneath part of one of the houses which fronted onto Vicarage Street. Housing of this type, in a narrow 'court' leading off from and behind another street, was quite common in Barnstaple in the last century. Contemporary maps and the Census refer to other examples such as Baker's Court, off the High Street at around number 85, and Pengelly's Court, near 109 High Street.[8] Neither of these exist today, their sites having been amalgamated into those of their High Street neighbours. However one such street has survived, having been in the ownership of the Dymond family for many years and situated at the rear of their former fishmongers in Boutport Street. This is Somerset Place, which although now 'gentrified' and altered at its far end where there has been some demolition to create an additional access, gives an idea of this closely built, small housing.

Unlike the other Derby streets, Boden's Row had no rear yards or gardens, just what appear on contemporary maps to be narrow alleyways running down the backs of the houses.[9] Rev.Boggis, writing in 1915 after the street had been altered, describes the houses as being *of the meanest description... the space is too narrow to allow a footpath on either side.*[10] In his 1956 report for the Compulsory Purchase Order and the slum clearance programme W. Rodgers, the Chief Sanitary Inspector of Barnstaple Town Council described the street thus,

> *the houses are in a double row, the whole street, which is a cul-de-sac, being barely fourteen feet wide. The houses are built of rubble to first floor height and above this, lath and plaster only four and a half inches thick. Rising and penetrating dampness is common... they are all wretchedly lit and ventilated, the majority have no proper food stores...(and)..no internal water supply.*[11]

The development of the rest of the area for housing followed fast on the heels of Boden building his factory. The deeds of a property at Union Street start with an indenture dated 25th. May 1826 for the lot of ground B45.[12] This is between Benjamin Baller the younger of Barnstaple, glover; Thomas Garland of Worcester, glover; and Elizabeth his wife, on the one part and Edward Richards Roberts; John Saunders; and Charles Roberts on the other. The condition was that within two years they would erect a,

> *...good and substantial dwelling house and that they would also bear one moiety of the expense of making a common sewer and paving the street in front of the said house and cause the same to be done before Michaelmas and from thence keep street clean and free from dirt stones rubbish or other thing or things which may inconvenience or annoy the passengers going through the same or the occupiers of the dwelling*

houses adjoining thereto.

Another, for 3 lots, B30,31,&32, naming Benjamin Baller and John Ulph, states that on each should be built *a good and substantial dwelling house... of stone or brick and cover the same on both sides with good slates.*[13] The houses in these three later Derby streets were larger and better built than their earlier neighbours in Boden's Row and had the benefit of paving to the front and of rear yards.

Similar indentures from the same time exist for lots of ground numbered C32 and C33 between Benjamin Baller and John Lake, glover, which would become part of Reform Street.[14] Neither Reform Street or Union Street are mentioned by name in the early indentures. The plots are identified by number and by their position on a diagram showing the three proposed streets on land known as Gaydon's Ground and/or Vicarage Meadow. A set of deeds for two properties in Newington Street has no such plan but refers to plot numbers 11 and 12 *in a new street called Newington Street near Vicarage Lane in Barnstaple.*[15]

Newington Street appears to have been so called after the old name for the area. A lease of 1882 refers to *lands at Newington in Barnstaple*[16], and the Newington Inn and Newington Cottage once stood just up from its junction with Vicarage Street on the site where the Miller Institute was eventually built. The grounds of the Institute itself continued to be known by their original name of Newington Gardens.

In his account of the parish of St. Mary Magdalene Rev. Boggis states that Union Street was so named as it formed the road which joined Derby with Gaydon Street, so formalising the unofficial footpath used by pedestrians and ending the long detour around to Bear Street for wheeled traffic. However a Journal report of July 1843 bemoaning the lack of such a proper way to Bear Street refers to Union Street by name and so Boggis' theory cannot therefore be correct.[17] It seems more probable that Union Street, like its neighbour Reform Street, was named in honour of the Reform Act of 1832, the celebrations of which, as noted in Chapter 2, were marked by the carrying of *a very beautiful lace flag* by the Derby Lace Factory operatives which bore the legend 'Union and Reform'.[18]

Princes Street, on the other side of Vicarage Street, was also being built at this time on land known as Prior's Bear. An indenture of 24th. June 1826 names Richard Gilbert, builder, Simon Bishop, carpenter, and Charles Roberts, gentleman.[19] According to Boggis again, this land was previously called Princes Meadow being owned by the Reverend William Prince. He was latterly Vicar of Landkey, but prior to this lived at Vicarage Lawn House.[20] No other reference has been found to prove or disprove this.

Higher and Lower Maudlin Streets reflect the name of the old priory of St Mary Magdalene once situated nearby (Maudlin is a corruption of Magdalene). The parish church was also named St Mary Magdalene to perpetuate this connection with the past.

The next indenture for lot B45 in Union Street, also from 1826, names John Musslewhit and William Gaydon, trustees of the Union and Friendly Society, and John Jones, schoolmaster and trustee, and refers back to that of 25th. May 1826 stating that this *did direct limit and appoint demise and lease unto the said Thomas Garland all that lot of ground Number B45 part of a certain field called Gaydon's Ground otherwise Vicarage Meadow.*[21] This was for a period of 2000 years at a yearly rent of £2 and five shillings and a loan of £50 was granted for the purpose. A long lease such as this was tantamount to a freehold nowadays but was so phrased due to the restriction on ownership of property before the reforms of the mid nineteenth century.[22]

W.F.Gardiner, writing in 1897, acknowledged the *immense value to the townspeople and particularly to the artisan classes* of these friendly societies in that they had *stimulated building enterprise, and encouraged thrift, and have enabled scores of working men to acquire the freehold of their leases.*[23] He credited the formation of the Union and Friendly Society in 1823 to Thomas Garland, whose grandson remained secretary of the same (renamed the New Union and Tradesman's Friendly Society) at the time of his writing. However, the rate books available for 1840, 1841 and 1879 show which of the houses in Derby were inhabited by owner/occupiers and, not surprisingly, these numbered relatively few.

There are deeds available in North Devon Record Office for several of the properties which were eventually bought by the Council under slum clearance and the road building programme. Although these only cover relatively few houses, all of the streets are represented once or twice and show their history from the initial building plot stage through the many changes of ownership down the years. Due to the numerous transactions and both the accumulation of and selling off of property the deed packets themselves are in some cases jumbled and partial. However what they do show is the variety of people who had property interests in Derby at any one time and that some houses were sold on at frequent intervals with sitting tenants.

Many well-known and moneyed Barnstaple family names can be picked out as landlords. These included John Westacott, shipbuilder in 1838[24]; James Maldram, woolstapler in 1870[25]; Emma Eliza Perrin and Arthur Frederick Seldon, solicitor in 1900[26]; John Cater, builder in 1907[27]; and Helena Minnie Petter in 1911.[28]

As stated above, tracing the history of the ownership of a particular house can be difficult and is further complicated as the houses appear to have been re-numbered for some reason in the late 1890s. This is shown by another agreement, dated 1st. January 1896 between William Curtis and Frederick Thomas Barwick, and which includes the schedule reproduced below,[29]

	Tenants	(Union Street)	Rent (yearly)
2&3	Mrs Found		
(now 1&2)	& Mrs Hill	"	£2. 2. 0
7	Mr Parkhouse	"	£1. 5. 0
(now 6)			
11&12	William Ovey	"	
(now 10&11)	& Mrs Ovey	"	£1.15.0
			£5. 2.0

Other evidence which discounts the 'workers' housing' theory can be drawn from the various trade directories published and the Census returns for the period. Billings 1857 Directory lists for example in its trade section,[30]

Thomas Dalling	master mariner	New Buildings
John Blackmore	boot & shoe maker	Boden's Row
Elizabeth Buzzacott	shopkeeper	Princes Street
William Gammon	builder	Princes Street
Nicholas Jones	cabinet maker	Newington St.
J Phillips	painter and glazier	Higher Maudlin St
John Rude	currier	Higher Maudlin St
John Symons	tailor	Lower Maudlin St

–to give just a few instances of the variety of occupations followed by those who lived in Derby.

Some of these supplemented their income further by providing accommodation for others. On reading the 1851 Census it appears that the properties of Elizabeth Buzzacott, described as a grocer, and her neighbour, a bread baker, must have been relatively large as both appear to have housed five separate households containing up to three generations. Half of these were either headed by, or contained young unmarried mothers working as lace menders. The 1840 rate book shows John Rude as the owner of two houses in Higher Maudlin Street and the owner/occupier of one in Vicarage Street. Another who accumulated property was John Stoyle, a builder, who built and lived in his own house in Princes Street. By 1855 he owned 4 houses in Princes Street and 2 in Newington Street, and by 1862 a cottage in Vicarage Street had been added to his family's holding.[31] In 1879 the tenants of the Newington Street houses were Charlotte Glanville and Richard Prust, both of whom had family connections with Derby Lace Factory.[32]

However, only a very small proportion of the housing at Derby appears to have been directly associated with the factory at any stage in its history. John Miller is given as the owner of 3 houses in Union Street, 2 in Lower

Maudlin Street and 2 in Princes Street in 1840, as well as his own home at Gorwell, the factory and buildings at Little Derby, and of land at Goosely Marsh, the site of the Newport/ Bishops Tawton lace factory.[33] The Derby houses were occupied by William Fisher, William Bragg and Samuel Geen, John Cockram and John Glover, and William Hole and Thomas Woodward respectively. The 1841 Census shows that all were employed at the factory.

Somewhat surprisingly, however, John Boden is still named as the owner of land at Stoney Bridge and indeed of the factory itself with buildings at Little Derby in the undated 1830s rate book, although John Miller is given as the occupier of both.[34] (Unless this is an error, it would appear that John Miller did not buy the business outright when he took it over in 1828.)

The rate books confirm that only from the late 1870s did the Miller Brothers themselves have any substantial property holding in Derby, and that this consisted only of 14 houses and one garden, being concentrated on the factory side of Princes Street.[35] They are not listed as owning any property in Derby outside Princes Street by 1879.

PUBLIC HOUSES

Those who knew Derby will remember that there were three public houses, The Union Inn which, as said before, survives today; the Curriers Arms, on the corner of Lower Maudlin Street; and the Carpenters Arms, which eventually formed the corner between Corser Street and Reform Street and at which Dai Morgan, a noted boxer, was publican in its later days. What may seem surprising today is that during the last century these premises were also used to hold inquests as well as other formal public meetings. The inquest into the death of George Cockram, the 4 year old son of a lace twister, was held at the Union Inn, near his own home in Princes Street, in May 1855,[36] whilst that of Henry Dunsford of Union Street was held at the Curriers Arms in 1853.[37]

In May 1855 John Nicholls, a lace twister at the Derby works, applied for and obtained a provisional transfer of the licence of the Carpenters Arms. The account printed in the Journal and the editor's subsequent comment gives us a flavour, doubtless biased, of how these premises were regarded by the town 'elders'. Nicholls was told by the Bench that *on the next licensing day they should expect the applicant to produce a certificate from his employer, Mr Miller – the eighteen persons whose names he had presented today were unknown to them.*[38] Obviously they were exercising caution as to who should be granted the licence on a permanent basis. The editor's comment more than underlined this and he certainly took the opportunity to vent his spleen on the subject when he added,

> *...we hope the magistrates will be exceedingly cautious as to the party whom they licence to conduct this house. From past experience, we are assured that the various persons who have successively occupied the "Carpenters Arms" have, by the irregularities they have permitted therein, contributed more to demoralise society in that locality than all the other vicious agencies combined. Persons of every age and sex – men, women, and children have been actors in the orgies within, and common decency has been outraged by the gross improprieties perpetrated without. If the premises must needs be continued for the sale of beer etc it is of the utmost importance to the interests of morality that they should be occupied by persons whose character and recommendations will be a guarantee against the repetition of such degrading and disgusting scenes as those we have referred to.*

Originally there was also the Newington Inn, on the site of the Miller Institute opposite the factory. A Journal report from June 1827 records that *Thomas Steed Domingo, a lad who is employed at Messrs Boden's factory,* was committed to trial *for stabbing Henry Seldon, son of Thomas Seldon in the Newington Arms* [sic] *...with an instrument something like a sharpened screwdriver with which some part of the machinery is adjusted.*[39]

In addition to these 'formal' premises there were also a large number of beer shops which appear to have operated from private houses, much as the 'shops' that some ran from their front rooms which sold milk and a few groceries. Gardiner mentions the demise of the Newington Inn but also recalls that,[40]

> ...*many houses have had short spells in the same line since 1837 . At the top of Higher Maudlin Street was the Dolphin, number 9 in the same street was known as the Victoria, number 34 as the King William, and number 36 was the Half Moon. In Lower Maudlin Street was the North Star, chiefly remembered on account of the doggerel which appeared on its three-cornered sign:—*

> *Step in and take a glass of my good liquor*
> *T'will help you up the hill the quicker*
> *If you tarry here till night comes on*
> *The North Star will guide you home*

Presumably this would be your final port of call before the long hill up to Goodleigh. Gardiner also recounts that Vicarage Street housed a licensed house, Princes Street had a beer house, whilst in Gaydon Street was the Mason's Arms *now disappeared, to say nothing of John Lake's cider shop in Reform Street.* Billings Directory of 1857 records John Lake, cider seller, of Reform Row whilst the Census of 1871 shows him as a glove cutter. The 1881 Census records Thomas Pedlar, a maltster, living in Princes Street, who might have operated the beer house referred to by Gardiner. That of 1891 shows Charles Garland, builder, living at 9 Higher Maudlin Street and at number 36 Elizabeth Garland, widow, living on her own means. As the licensee of the Curriers Arms, on the corner of Lower Maudlin and Vicarage Street, was John Garland this may point to a family connection in the licensed trade.

7

The two photographs

TWO PHOTOGRAPHS

As stated earlier, no records of the factory survive from before the Second World War and so our only evidence about the workers and their families has to be found in the Census records – available from 1841 – 1891 inclusive, the parish registers of St Mary Magdalene church, and from the North Devon Journal for those involved in court cases, or other news stories.

However, a small selection of photographs do survive and two of these are captioned so that we can at least put faces to names, some of which have become very familiar through their long or eventful association with the firm and have been mentioned earlier in this account.[1] The mounts and frames have been renewed in recent years and this may account for some errors in transcription which become evident when checking through the details.

One photograph is captioned 'The Mill Staff 1883' and the other 'The Wareroom Staff 1883'. Luckily this date is relatively close to that of a Census year, 1881, and even more fortunately this is one Census which is fully indexed for Devon. With some time spent checking through the alphabetical listings, matching ages and occupations of potential namesakes, we can be fairly sure of the identities behind the likenesses. Reference then to the full enumerator's entry gives us their address and details about themselves and the other members of their household such as their occupation and place of birth. In this way we can find out more about their background and, if further cross-reference across the Census years is possible, can show some of the changes which took place in a particular person's lifetime.

No Christian names are given in full on the captions, only the initial letter, and in the case of some of the older women they are merely described as 'Mrs' and then their surname. The photographs themselves help to a certain extent in choosing between like-named people of different ages, if the person looks particularly young or particularly old. However it can be difficult to place ages between these, especially given the similarity in hairstyle of the women, and facial hair, often 'full sets' of whiskers, of the men. Should any descendants disagree with the identities so allocated to their ancestors we apologise and are willing to stand corrected.

It is difficult to gain much information from the photographs themselves. They were both taken outside the factory in the same location. The clothes are, as expected for the era, very dark in colour which makes detail hard to pick out. This is not a fault of the prints themselves which are excellent in quality. The men appear to be dressed for the most part in fairly long, almost three quarter length in some cases, single breasted jackets worn open to show a waistcoat underneath and in many cases a watch chain. Some are wearing hats of a bowler or style in between that and a top hat. One has his hat, which looks more of a top hat shape, on a box or something on the floor behind him. Perhaps he had shied away from wearing it in the photograph after all. Another is holding his hat on his lap, perhaps because he is in the presence of the ladies. Their footwear does not appear to be particularly heavy duty in style although this is hard to judge, but may further signify that the men have come in their 'Sunday best' for the photograph to be taken.

The women are also dressed similarly, in close fitting single breasted jackets or overblouses from under which varying quantities of white lace collar are showing. The jackets are fastened with many small buttons, some white or pearl, others self-coloured. One woman wears a head-dress of some kind, the others all have their hair parted in the middle, except one who has a fringe, and fastened behind them somehow. One woman wears a medallion and a watch chain. All wear long skirts, which could be of a chintz type material, and which are heavily ruched and frilled. Consequently only the tips of shoes can be seen and then only in very few cases. Again this is likely to be 'Sunday best' dress for the photograph.

The year 1883 was one of the many in which a *wayzgoose* was held for the factory workers, together with those of Messrs. Baylis & Co. of Pilton Glove Factory.[2] The Journal records that,

> *a special train was chartered, and a very large company was taken to Plymouth where the workpeople spent a very enjoyable time. By a liberal arrangement with the Railway Company, on payment of a nominal sum, a number of the excursionists remained in Plymouth for two or three days.*

It is possible that these photographs were taken to commemorate the event but unfortunately there is no mention of such pictures having been taken to mark the occasion.

THE MILL STAFF 1883

Seated to the furthest left of the photograph is Louis Watts. Louis is recorded in the 1881 Census as a machine fitter, living just outside the Derby area at 1 Grosvenor Terrace with his parents and brothers and sister. His father Philip is listed as a house painter, his mother Annie a dressmaker originally from Marylebone, London. He also had a sister Edith, aged 17, a pupil teacher, and two brothers Frank and William aged 13 and 9, both scholars. At the time of the Census they had a visitor staying with them, Harriet Alexander, an

Louis Watts

upholsteress from St. Pancras, probably an old friend of Annie.

The marriage register of St.Mary Magdalene church records that on the 22nd. April 1886 Louis Watts, then aged 25 and an engineer, married Evelyn Adelaide Rawle.[3] His wife's father is given as Richard Rawle, timber merchant of the parish, and was possibly connected with the builders' merchants, Rawle, Gammon and Baker, which survives today.

I or J (?) Bennett

Next in line is named as 'F.Bennett'. There is only one possible candidate with this initial in the 1881 Census index, Frederick T. Bennett, whose father Thomas worked at the factory and is elsewhere in the photograph. Frederick was aged 14 in the 1881 Census and recorded as a pupil teacher. Although it is difficult to ascertain age by the photograph it is unlikely that the fresh face of a 16 or 17 year old lurks underneath that bushy handlebar moustache. It is more likely that in transcription an 'I' or a 'J' has been mistaken for an 'F' and that this could be either Isaac Bennett, an engine fitter, of Lower Maudlin Street, or John Bennett, 'foreman, iron foundry' of Grosvenor Street.

Isaac, aged 39 in 1881, lived with his parents and was the son of a lace operative, also named Isaac who himself had originated from Torrington. The 1861 Census shows the family again living in Lower Maudlin Street with Isaac, a turner and fitter and eldest of seven children, living at home. His younger brother George also worked at the factory as a lace operative's assistant.

John, aged 45 in 1881, appears to have been a brother to Thomas Bennett, who also appears in the photograph, as both hailed from Linkinhorn, Cornwall. As well as working in the foundry he is also described as being a 'pensioner (seaman)'. His wife Emma's place of birth, Stonehouse, Plymouth, and those of her children, Robert and Emma Smith, born in Torpoint and Stonehouse, both naval ports, suggest that the couple met during his years as a sailor. They also had two other children, Thomas and Frederick Bennett, both born in Barnstaple.

A transcription error and also poor positioning of the caption causes some confusion in establishing the identity of the next man standing to the right. The caption reads 'F. Fear' of which name there is no trace in the 1881 Census. However there is John Fear aged 54, an engineer in the factory.

John Fear

John Fear appears in each Census from 1841 – 1881 inclusive, living in Vicarage Street from 1861 and variously described as a machinist, a lace machine builder, and as an engineer in the lace factory. He was the son of Thomas Fear, originally a lace twister and latterly a storekeeper at the factory who had been retired with a pension by John May Miller as mentioned earlier. In the 1841 Census, not only was Thomas still a lace twister at the age of 69 but he also had an eight year old daughter Eliza! John at this time was aged 10. Other members of the household were wife Sarah, twenty years his junior at 49, and daughters Charlotte, 25, a dressmaker, and Mary, 15. The large age gap between Thomas and his wife, and the young ages of the children may indicate that Thomas had married a young widow with a family, which was fairly common at that time.

By 1851 the family had moved from Boden's Row to New Buildings, and although the ages given for Thomas and his wife are consistent, those of the children are more disparate. John himself was now 24, and a lace machine maker, and had seemingly gained four years, whereas sister Eliza was now twenty, and a dressmaker. Mary's age, 26, is about right allowing for the varying months when the Census could have been taken.

In the 1861 Census John was 31 years of age and had married and moved to Vicarage Street. He was described somewhat misleadingly as a machinist and had a wife, Mary, 32, and sons Alfred and Francis James, both scholars, aged 6 and 3.

In 1871 he was once more described as a lace machine builder of Vicarage Street, aged 47 years, and his wife Mary as 42. Their children are listed as Francis, aged 13, Herbert, 8, and Claude, 5.

By the time of the Census nearest to the photograph (1881) John was described as an engineer in the lace factory, aged 54, which tallies with the age given in the 1851 census. Mary was now 52 and of the children only Herbert, now 18 and a printer compositor, remained at home.

The next person seated appears, allowing for the misplaced caption, to be George Viney. George lived at Higher Maudlin Street in 1881 with his wife Mary, from Exeter, and six children, all born in Barnstaple, aged between 4 and 14. The eldest, Laura, was a dressmaker, the others scholars. George himself was then aged 37 and described as a foreman in the lace factory. He originated from Cullompton, a woollen town, as did some other Derby residents,

George Viney

perhaps indicating movement between textile areas to find work in the new, growing industries as the traditional ones declined.

George appears in the 1861 Census as a 17 year old smith, then living in Vicarage Street with his widowed mother Charlotte Glanville, a 43 year old washerwoman. Charlotte, and George's younger brother and sister, William, an errand boy and Elizabeth, a scholar, were also born in Cullompton but his half brother Thomas Glanville had been born in Exeter two years previously. By 1871 Charlotte had moved to Newington Street and was working as a lace mender, William was now an engine fitter, possibly also at the factory. Charlotte was living alone in Newington Street in 1881

The 1891 Census finds George still working as a foreman at the factory, but now living at 13 Fort Street, then newly built. Laura still lived with her parents, now a milliner and dressmaker. Albert was now an apprentice lace machine fitter whilst Beatrice aged 14 had no occupation listed. Another machine fitter's apprentice, Samuel Vine or Hine, aged 15, boarded with them which would have helped the family budget now that they were living in a more expensive area.

George's marriage does not appear in the St. Mary Magdalene records – his wife probably came from a different parish – but those of his daughters do. Maud Henrietta married Harry Robert Harding, a school master from St Johns, Hackney, on the 28th. December 1896; Edith married Harry Hawkins four years later on Boxing Day, and Beatrice the youngest, married John Green in April 1906.

Standing further to the right of George is Samuel Geen. He appears in each Census from 1841-1881 firstly recorded as a smith living in Union Street, then as a whitesmith in Newington Street and finally in Vicarage Street where he continued to live being described latterly as a machinist. In 1883 he would have been 69 and was obviously still in physical work at an age which would not be expected today. In 1841 he had a wife, Jane, and three children, Charles, 5, Elizabeth, 4, and Samuel, aged 1. By 1851 Charles had become a smith's apprentice, and three more children, Mary Jane, 8, James, 3, and Martha, 5 months, had been added to the family. William, 8, had joined by 1861, and Charles had left home. Elizabeth was now a waistcoat maker, and Mary a dressmaker. Samuel and James had followed in their father's footsteps, one a turner and fitter, the other a whitesmith.

Samuel Geen

His son Charles is the next

man seated in the photograph, also described as a 'machinist – lace factory'. In 1881 he lived in Newington Street and was aged 45. The Census entries for the family are interesting in that we can see where they have moved away from and back to Derby. Martha, Charles' wife, originated from Ashburton, but their eldest child, Emily, was born in Barnstaple, perhaps implying that Martha had moved to North Devon before her marriage. The

Charles Geen

next children, Charles, 19, a printer and compositor, and Henry, a scholar, were also born in Barnstaple, but then come Elizabeth, born in Woolaston, Gloucestershire, and Percy, 5, born in Chepstow. The youngest child, Martha, aged 5 months, was born in Barnstaple. The family could then have moved back to the town at any time during the previous five years.

Another thing that can be learned from these entries is that the facts given in any one year are not necessarily accurate as discrepancies can be found when entries are compared over the years. The 1881 entry shows signs of alteration and Emily's age appears to be 27, whilst Henry's has been changed from 9 or 7 to 11. Checking back to 1861, when Charles and Martha were living in Reform Street, Emily, their only child at this time, was 2 years old which would make her 22 in the 1881 Census above. Inconsistencies in ages are quite common to find when looking through Census records and can have implications for those doing family history who might only refer to one set, for example to establish the year of birth of an ancestor by which to consult parish records.

Charles and his family are missing from Derby in the 1871 Census, which correlates with the beginning of the period of absence during which Elizabeth and Percy were born. This appears to confirm that Henry, born in Barnstaple, was aged 11 in the 1881 Census, as he would have been born in around 1870.

By 1891 only Percy, then 15 and a printer compositor, and Amattie (Martha) aged 10, a scholar, were still at home with their parents. Charles' occupation was given as 'machine fitter' and the family had moved to Vicarage Lawn, built on the former grounds of the large house of the same name once lived in and run as a school by William Snow.

Charles Huntington Geen died at the grand old age of 81 at Higher Maudlin Street. He was buried on April 29th 1917 in Bear Street Cemetery and his funeral was *attended with many manifestations of sympathy and regret*.[4] His widow Martha, with whom he *would have celebrated their diamond wedding* in a few months, *was unable to attend owing to disposition*. She herself died in the following January and was buried with her husband.[5] *Other relatives prevented from attending through long distances* were Henry, then living in Ireland, and *the youngest daughter*, Mrs E.H. Widlake, presumably Amattie, who was in Canada. Son Charles and two other daughters were present, and the bearers were *fellow workmen at the Derby Lace Factory at which deceased had been employed for a long series of years*.

Sitting beside Charles Geen is Thomas Bennett. Thomas came originally from Linkinhorn, a small village in Cornwall. He is recorded in 1881 as being aged 39 and living at 12 Richmond Street with his wife Eliza and children, Frederick aged 14, a pupil teacher, Albert, Edith, and John, all scholars, and three boarders, Mary Lethaby, a 29 year old dressmaker, Charlotte Braddon, a 21 year old machinist and James McConnell, a 47 year old Scottish general labourer and pensioner.

Thomas is described as a stoker in the lace factory, an occupation that he might have learned in an early steam powered ship. His wife Eliza was born in Barnstaple, Frederick was born in Gibraltar, and Albert was born at sea –

Thomas Bennett

perhaps as the couple were returning to Britain? The parents were in Barnstaple for the births of Edith and John.

Standing beside Samuel Geen is Richard Pearce who would have been 48 in the year the photograph was taken. He appears in four Census records as living in Higher Maudlin Street, his occupation varying from smith, machine fitter, or as in the 1881 Census, 'engine fitter at lace works'. At this time he had a wife

Richard Pearce

Mary and two children, Herbert, 16, a silversmith, and Charles, 7, a scholar, living with him. Unusually for Derby they are also recorded as having a general servant, Elizabeth Toute, a 13 year old girl from Swimbridge. The inhabitants of Higher and Lower Maudlin Street tended, however, to be more of the skilled and artisan class than their neighbours in Princes Street or in the majority of the streets on the other side of Vicarage Street.

Thomas Sutton

Next to Richard Pearce stands Thomas Sutton who, with his wife Elizabeth, shared a house in Higher Maudlin Street with another worker at the factory, Robert Jordan, and his family. Thomas originally came from Macclesfield and was at this time a watchman although previously he had worked as an engine driver at the Derby works.

George Leworthy, next in line,

George Leworthy

lived at some distance from the
works by Derby standards, outside the parish and across the
river Yeo on Pilton Quay. He was
a machine fitter and was married
with three children, Hannah,
Beatrice and Marian. Richard
Bray, a young blacksmith from
his wife Anne's home parish of
Berrynarbor, boarded with the
family.

William Quick

The man with the springer
spaniel is William Quick who, like
Samuel Geen, was still working in
the factory as a labourer at the age of 69. His wife
Richard, a name not now given to female children, and
their daughter Elizabeth, a machinist, lived with him in
Princes Street at the time of the 1881 Census, in one of
the houses then owned by Miller Brothers. In the 1891
Census he is described as a widower and a 'retired coach-
man'. Elizabeth was still living with him and was a silk
winder, and they had a young cotton winder, Elizabeth
Offield, staying with them on Census night.

To the side of William Quick stands Thomas Seager,
another 'old timer', quite literally in his case as he was
timekeeper at the factory and is seen here in his 71st
year. In 1881 he lived at 10 New Buildings with his
brother William and both origi-
nated from Tiverton. William is
described as a 'widowed wool
comber, out of employ', another
example of the movement between
textile areas and fields of employment. Thomas is described as
married but no wife is listed. The
household was completed with
William's unmarried daughter
Mary, a vest maker, and two
elderly boarders. Ten years earlier
Thomas had been lodging with
John Moore, a lace twister of

Thomas Seager

Higher Maudlin Street and was then himself listed as gatekeeper at the factory.

The young man seated and wearing a bowler hat is George A. Beer, a machine fitter aged 18 in 1881 who originated from Plymouth. He lived in Princes Street with his wife Amelia, then aged 21 from Kings Ash, who is listed as a lace bobbin winder despite having a two week old daughter Ellen, born in Barnstaple. Like Charles Geen, George

George A Beer

and Amelia Beer had moved to Vicarage Lawn by the 1891 Census. Ellen was still a scholar, although now 14, and in addition there were Arthur and Minnie, also scholars, aged 8 and 6, and Maud, aged 2.

Robert Jordan, the remaining man standing, appears in each Census from 1861-1891, described as 'blacksmith', 'smith at the factory', or as 'engine smith at lace factory'. By 1881 he was widowed, his sons James, a tailor and Henry and Frederick, scholars, living with him. As already stated, the house was shared with Thomas Sutton, and the Census ten years previous shows a real houseful – both men and their wives, as well as six Jordan children aged from 17 down to one year old, Thomas's adult son Samuel, and grandson John aged 10.

Lastly, sitting at the extreme right of the photograph is Philip Gardiner. Philip can almost be said to be a true Derby boy having grown up and spent most of his life in the area, but he was actually born elsewhere. He appears in the 1841 Census, as a 12 year old boy living in Newington Street with his mother Ann, a charwoman, being the eldest of her six children then

Robert Jordan

recorded. By 1851 the family had moved to Princes Street, sister Mary had left home and Philip was now a lace twister, younger brother John was also working in the factory as a brass threader. This was the first Census to give places of birth and we can now see that Philip's mother hailed from East Buckland, the two eldest children were born in South Molton, the next two in Filleigh, and the youngest in Barnstaple – showing a slow

Philip Gardiner

migration towards the town. The 1861 Census finds him in Higher Maudlin Street aged 32, still a lace twister, with a wife, Fanny and children Gertrude 8, Caroline 4 and 1 year old Philip Henry. In 1871 he was described as an overlooker, living in Vicarage Street, as he was in 1881, near to the time of our photograph. By this time he and Fanny had five children all born in Barnstaple and aged between 18 and 9, and also a young boarder. The 1891 Census states that he was now retired, and two daughters, Mary Ellen aged 26 with no occupation, and Edith, 24, a music teacher, were living at home.

THE WAREROOM STAFF 1883
Sitting on the extreme left is Mary A. Nicholls, who is recorded in the 1881 Census living with her 3 year old grand daughter Annie, in Princes Street. Mary was then 58 and a widow working as a lace mender. Annie is described as a scholar despite her young age, and perhaps attended school while her grandmother worked, if this was a permanent living arrangement.

Ten years later Mary appears to have retired from the factory,

Mary Nicholls

86

as she has no occupation listed. Her son Arthur, 30, an iron moulder, and a lodger, Florence Buckingham, an 18 year old lace mender, shared her household. The 1861 Census shows Mary and her family to have been highly dependent on the factory for their livelihood. Mary's husband John, then 43, was a lace twister, and their four eldest children aged from 16 down to 9 were respectively Ann, a lace mender, John, lace machine boy, Elizabeth, lace

Mrs Balment

bobbin winder, and William, threading boy. Mary also had Thomas, aged 5, Arthur, 3, and Lucy, 8 months, to look after. John Nicholls held the licence of the Carpenters Arms for a brief period in the mid 1850s and it was this application which provoked the response from the editor of the Journal quoted earlier.

Next to Mary Nicholls is a 'Mrs. Balment'. The most likely candidate from the 1881 Census is Annie Balment, aged 38, who was married to Thomas Balment, a cordwainer of Pilton Quay. Annie had five children under 11 at this time, the youngest being only 18 and 3 months old, and her aged father-in-law lived with them too. It is therefore unlikely that this is the same woman and that this woman was married, and therefore changed her name to Balment, between the Census and the date of the photograph if she was living in Devon at that date.

Standing with her hand on Mrs. Balment's shoulder, wearing the head dress, is a woman named as 'Mrs Passmore'. This is Ann Passmore, aged 67 in the 1881 Census and so nearing seventy here, and listed as a 'female overlooker in a lace factory'. She was then living in

Ann Passmore

Vicarage Street with her husband Edward, a tailor, and her grand daughter Minnie Pedlar, a 16 year old draper's assistant.

Ann first appears in the 1841 Census, aged 25, living in Lower Maudlin Street with her husband and son William, then 3, with a fifteen year old girl, Maria Dunsford, sharing the household. The 1861 Census shows that this was probably Ann's younger sister, as at this time Ann's household contained eight people – sons William and John were now smiths, and George was a coachbuilder, and Ann's parents William Dunsford, a lace twister, and Agnes his wife had joined the household. Emily Pidler (sic) a five year old scholar, and probable sister of the above-mentioned Minnie, lived with her too. This entry confirms that Ann was a member of the Dunsford family, who originated from Tiverton, and many of whose members remained and worked in Derby.

Arthur Stevens

On Census night in 1841 Ann's daughter Mary, aged 5, was staying with William and Agnes Dunsford and their son Henry, aged 20, in Boden's Row. William Dunsford was then a wool comber and Henry a lace twister. Another William Dunsford, also a lace twister, aged 25 and presumably Ann's twin, lived along the street with his wife Fanny. We shall meet Henry Dunsford's daughter later in this chapter.

The man sitting on the stool holding his hat, perhaps as a mark of respect for the ladies, is Arthur Packer Stevens, who was still employed at the factory as an accountant at the age of 74. In a Journal report of a court case involving an employee in April 1843 he stated that he had then worked in the factory for twelve years, which would mean that he had started there in 1831, and so by this time had given 52 years service. He married Mary Ann Pyke in Barnstaple in June 1833.[6] A report of the Revising Barristers Court from October 1836, which considered objections to those people who had claimed the right to vote, reads as follows.[7]

> *Arthur Packer Stevens was opposed on the ground that the house he occupied was not of the yearly value of 10/-. The claimant deposed that he had been in occupation of the house for the period required by law; that he had been repeatedly offered 10/- a year for the same, and considered it would fetch that, or even a much higher rent, if he chose to let it which he had never felt disposed to do so – Vote allowed.*

He appears in the Census as living at New Buildings from 1841-1861, listed as either an accountant or a 'clerk in lace factory'. A Poor Rate Book of 1840 shows him as the owner occupier of his house.[8] The 1841 entry is very difficult to read but he was then aged 30 and had his wife Mary, three young sons, John, William and Henry, and an infant daughter also called Mary. By 1851 John had become a grocer's apprentice, while William and Henry had followed their father into office work – one as a scrivener, the other as a solicitor's writing clerk. Mary was then a scholar, as was her younger brother George. By 1861 she was the only one of the children still living at home, when she had no occupation listed.

Arthur Stevens is missing from Derby in 1871. Whether this move was forced by ill health or perhaps became a cause of it is unknown. He was stated as being too ill to attend the celebration held by the lace workers in honour of John May Miller in 1872.[9] He is no longer listed as the owner of any house in New Buildings in the 1879 Poor Rate Book so for some reason he had sold his property there.[10] The 1881 Census finds him in Well Street, not as a householder but lodging with William Curtis and his wife Sarah. There is no record of his wife Mary, and there is no indication that he was related to the Curtises in any way. William Curtis was a mason and many of the households in Well Street were at this time headed by skilled workers and supplemented by one or two lodgers.

Standing next to Arthur Stevens is Catherine E. Wollacott [sic]. Catherine, aged 17 in 1881, lived with her widowed mother Ann, a charwoman, and her sister Mary Jane, then 13, at Boden's Row. Both sisters are described as lace menders. Catherine was born at Budehaven and her sister at Pilton. Going back to 1871 we find her father Henry alive, then aged 39, and an elder sister Eliza, three years her senior and also born at Bude.

Catherine Wollacott

An entry in St. Mary Magdalene marriage register records that Catherine Elizabeth Woollacott [sic] aged 20, daughter of Henry, and living in Newington Street, was married that day to Thomas Ridd aged 22, a mason of Reform Street.[11] The 1891 Census finds them living in Vicarage Street. Thomas, now 28, was working as a tanner, whilst Catherine, 27, had no occupation, and they had a six year old son Alfred, a scholar. Catherine's mother Ann, and sister Mary Jane, now a lace twister, shared the household.

Catherine is resting her hand on the shoulder of one of two women identified as 'E. Prust', the other being second on the right seated in front of the bearded man. One of these is Elizabeth Prust, recorded in 1881 as a 47 year old wareroom woman living with her widowed mother, also Elizabeth, aged 77, in Newington Street. As neither of the women in the photograph look old enough to be the mother, it is probable that one of them is in fact Susan Prust, Elizabeth's elder sister by two years, who shared the household and is recorded as 'forewoman in the factory'. There is certainly a marked resemblance between the two women, who were still living together and working at the factory in 1891.

One of the Prust sisters

90

The family appears in the Census from 1841 onwards. At that time they were living in one of the cottages at Rawleigh. Peter Prust was a blacksmith and, although it is not made explicit in the Census it is reasonable to assume that he worked at the Rawleigh Lace Factory as they lived in one of the factory properties. Neither his wife, nor any of his 4 children, have any occupation recorded. By 1851 the family were living in Newington Street

the other Prust sister

and so, unlike many of Heathcoat's workers, Peter Prust had decided to stay in his native North Devon rather than make the move to Tiverton. He originated in Bideford and his wife from Roborough. Peter then aged 53 was still a smith, probably now at the Derby factory, his wife Elizabeth again had no occupation listed. Susan, 21, was a bobbin winder, Elizabeth, 19, a lace mender. John, 16, an errand boy, and Richard, 12, a scholar. Although there is no direct reference in any of the Census years to his being employed at the Derby works it is highly probable that he was, especially given his daughters' employment there and his past experience at Rawleigh. By 1861 only Susan remained at home with her parents, being then aged 30 and still working at the factory, but in 1871 Elizabeth had returned to join her at home, and they were both described as 'lace workers'. Peter was then 72 and his wife 70.

St. Mary Magdalene parish records show that Susannah Prust was buried on 14th. September 1912 in the cemetery at Bear Street and Elizabeth on the 15th. January 1922.[12] Their headstone also commemorates their parents. A Journal report of 26th. September 1912 states that *the funeral took place on Saturday week of the late Miss Susan Prust, for 60 years employed at the Derby Lace Factory.*[13] Among the chief mourners were *Miss L Prust (sister) and Mr R Prust (brother)*, ie. Elizabeth and Richard, whilst one of those representing the factory was

91

W.M. Jones, who is also in this photograph.

The young woman with her hand on Miss Prust's other shoulder is Bessie Dunsford. Bessie's name crops up a number of times in the history of the Derby works and it is particularly pleasing to be able to put a face to her name.

Bessie Dunsford

Bessie first appears in the 1851 Census as a six week old baby. Her father Henry, a lace twister originally from Tiverton and brother of Ann Passmore, was then aged 31 and her mother, from Ilfracombe, was 26. Charlotte Harding, a young widowed lace mender lodged with the family in Union Street. Five children, all under ten, are named, the four older ones described as scholars – John, William, Elizabeth Ann, and Mary Jane. They were all born in Barnstaple.

In September 1842 Henry appeared in the North Devon Journal, along with William Harding, a painter, possibly Charlotte's husband, charged with the theft of apples from Mr Bryant of Pitt. They were fined 30/- each or three weeks in jail.[14] Henry obviously either could not, or would not, pay the fine as in December of that year he appeared in the Journal again – this time charged with assault by Samuel Elson, a smith at the lace factory. This incident had come about *on the 25th last.... They were at the Curriers Arms on Saturday night where complainant provoked defendant by throwing up to him that he had lately been in prison for stealing apples at Pitt orchards which had led to the assault.* On this occasion Henry was fined 2/6 plus expenses.[15]

However a far worse fate was to befall Henry, this time as the result of an incident in another Derby hostelry, the Newington Inn. He became involved in a drinking contest and *engaged to drink half-a-gallon of beer in eight minutes.* Apparently he often accepted bets of this sort, the prize on this occasion being 8d. After successfully drinking the quota in seven and a half minutes he was taken home *in a state of insensibility* by his companions, where Mary, used to him arriving home in such states, said for them to leave him on the stairs. They did so and, after checking him and finding him snoring, she went back to bed. Tragically, one of the children, eleven year old John, woke early the next morning and found his father dead and cold.

The inquest was held next day at the Curriers Arms, and the coroner and jury went to view the body at the house which they found *denuded of every comfort, and the family in a state of wretchedness, such as may be conceived better than described.* They returned a verdict of death by excessive drinking and gave their jurors' fees to Mary Dunsford and her now six children out of sympathy.[16]

In 1861 the family were still in Union Street, Mary was working as a lace mender and her two youngest children, Bessie and Martin, two years younger, were both 'winders of lace', as was Lizzie Ann. Mary Jane, the middle daughter, was entered as a scholar.

By 1871 Martin, now aged 18, had become a shoemaker, but the rest of the family continued to work in the factory – Mary still as a lace mender, Elizabeth as a silk winder, Mary Jane as a 'winder on lace mills', and Bessie as a bobbin winder.

The year 1874 saw the period of discontent at the Derby works involving the long 'lock out' of men and the court case brought against these men for conspiracy. As a result of these events Bessie was asked to work a lace machine. Her testimony in the report of the court case reads that,[17]

> ...*some years before she had worked in Messrs Miller's lace factory. Had been in America, and her master had sent her some money to come back to his employment. That was in 1873. Had worked in the factory ever since. Had been employed up to the 16th or 17th September at bobbin mending. She was then put on to working a lace machine. This work required study. Last Sunday week James Turner had called after her and said she was an "old horse marine" and told her to "change her clothes". Witness knew Turner's voice.*

There were others shouting out too whom she did not recognise. She had been called "old boy" and lots of other names. These cries annoyed her. She had been called out to on leaving work at Princes Street near the factory. On Friday week she had heard a disturbance near the factory. John Nicholls, George Peters, Charles China, and Mrs Peters were amongst the crowd. Mrs Peters came up to witness and asked if she was going to fight.

The dictionary defines a horse marine as 'a person quite out of his element' – the implication being then that Bessie was in a job which she should neither be in, or be expected to do properly.

Subsequently a scuffle had developed with Bessie being jostled and slightly hurt by the locked-out hands, but an argument and fight involving another worker, Robert Cure, who they accused of being a 'black sheep', took over. Bessie stated that she had not been called names since and that she was *still working at Messrs Millers' factory, at the work the men would be employed at. Unless there was a stop put to the row, she would be afraid to go to the factory when it got dark at night* because of threats that she and others would be *served out when it became dark at nights.*

Obviously Bessie continued at her machine as she is described as a lace twister in both the 1881 and 1891 Censuses. The animosity towards her did not end immediately though; another Journal report in July 1876 states that she had been abused by a drunken William Cawsey, who was not named as being involved in the original case. Again this consisted of being called after in the street and accused of wearing the *wrong clothes*, ie. being a woman doing a man's work. Susan Prust stood as a witness to this event.[18]

These reports indicate something of Bessie's strength of character, that she was not afraid to take on new challenges, having tried her luck in America, and in taking on a 'man's job' under difficult circumstances. It also says much about how the Millers valued her as they sent her the money to come home from America and then chose her to take on the new work. For Bessie's part, she stood as a witness in an extremely delicate court case involving many of her neighbours. However, this may not have been entirely to return a favour. Title deeds for a property at 36 Princes Street refer to an indenture made between Thomas Garland and John May, William Walter, and Alfred Henry Miller, dated 24th. July 1875.[19] The house where Bessie and her family were to live in Princes Street was one of the 14 the Millers owned in this street. Her brother Martin was also to live in a 'factory house' although, as a shoemaker, he did not appear to work for the factory once he became an adult. It is impossible to say whether Bessie and her brother were living in 'tied properties' at the time of the court case, but obviously, if this was the case, it could have had a bearing on her decision to stand as a witness.

In the 1881 Census Bessie is recorded as a lace twister, living in Princes Street, and then aged 31. Her mother Mary, a widow, worked as a lace mender, and her elder sisters Elizabeth and Mary were lace wheel winders. Her older brother John also lived with them and was described as a pensioner although only 39 years of age, so perhaps this referred to an army pension.

In April 1884 Bessie Marie Dunsford, aged 33, lace twister of Princes Street, married Edward Chambers, aged 28, a sailor in the Royal Navy, also of Princes Street, at St. Mary Magdalene Church.[20] Edward's father was a fruiterer in Joy Street.

However Bessie was included in the Dunsford household on Census night 1891, of which her sister Elizabeth, still a lace winder, was now the head. Her other sister Mary Jane, now with no occupation and described as an imbecile, along with her mother were its other members. Presumably Bessie's husband was at sea. Elizabeth and Bessie seem to have reduced their ages now that they had reached their forties, so that Mary now appears to be the eldest sister, and Bessie herself appears the same age as her younger brother Martin, who lived along the street with his wife and three children. Her older brother John was lodging with a family in Portland Buildings.

A few years later Bessie and Edward had their only

child, Frederick Edward, who was baptised at St Mary Magdalene Church on the 20th September 1894.[21]

Bessie was buried in Bear Street Cemetery on the 16th. September 1924 after a service at St Mary Magdalene Church. Her obituary notice, printed in the Journal a few days afterwards reads as follows,[22]

> *To the deep regret of a wide circle of friends the death took place at 30 Princes-street, Barnstaple on Friday of Mrs Bessie Maria Chambers, wife of Mr Edward Chambers. Deceased, who was aged 73 years, was a native of the town, in which she resided. Of a particularly bright and unselfish disposition, the late Mrs Chambers was held in the highest esteem. She bore with patience and fortitude a long illness. For 54 years she was an esteemed employee at the Derby Lace Factory, Barnstaple in the wheel and bobbin winding department. With the bereaved husband and only son (Mr F.E. Chambers) deep and sincere sympathy is felt.*

Lizzie Ann, her only surviving sister died in April 1926[23] while brother Martin had passed away in November 1912 aged 59, still living at 32 Princes Street.[24]

Sitting with a cat on her lap is a woman named as E. Harris. There are numerous women of this name recorded in the 1881 Census in various occupations, and once again it is very difficult to ascertain the age of this woman from the photograph. The most likely candidate would appear to be Eliza Harris, a widow aged 52 in 1881, living at Boden's Row with her unmarried children, Charles Pugsley, a 29 year old described as 'Late seaman, RN' and Mary Harris, aged 16. Both Eliza and her daughter were lace menders. It is possible that the woman in the photograph is in fact the daughter. As said already, it is difficult to place this woman's age, especially when she is compared on one hand with the Prust sisters, both in their late

Eliza Harris

96

forties/early fifties, and on the other hand Catherine Wollacott and Ann Dyer, in their late teens/early twenties. The marriage register of St. Mary Magdalene church records that on the 22nd of December 1863 Eliza Pugsley, widow of Boden's Row, daughter of John Burgess, cordwainer, married Edward Harris, labourer, also of Boden's Row.[25] By the 1871 Census she had been widowed again and was lodging with Susan Hill, mariner's wife and sister of Bessy Harding. She was at that time a lace mender and her daughter, a scholar, is named as Polly.

The parish registers show interesting entries in that Ann Burgess, widow of Vicarage Lane, Eliza's mother, is recorded as having a bastard daughter called Eliza baptised on the 18th. October 1826.[26] Sadly an eighteen month old child of this name, daughter of Ann, Derby, was buried on May 30th. 1828. However on the 17th. August of that year Ann Burgess had another bastard daughter, also named Eliza baptised. The year of birth for this child correlates with the age of the Eliza who grew up to become Eliza Harris.

Next to the right, seated is Hannah Paltridge. Hannah's family can be traced using the Census records over four generations and once again a pattern of movement between textile areas can be seen. In 1881 Hannah, aged 24, and her sister Esther, 17, were both listed as lace workers, and their younger brother James was a lace twister. They lived in Princes Street with their parents – John, their father being a fellmonger. Along the street lived their elder brother Henry, a lace twister, and his wife and children,

Going back to 1871, Hannah, her mother (also Hannah), and her brothers and sisters were living in the household of their uncle James Pinkstone, a travelling pedlar, in Boden's Row. Her grandmother Esther Pinkstone, aged 84, was on parish relief, and both she and James hailed originally from Nottingham. The elder

Hannah Paltridge

97

Hannah, Esther's daughter, was born in Tiverton and is described as a mariner's wife. Besides the young Hannah there were 19 year old Louisa, a silk drummer; Henry, 17, lace twister; Esther, 7, scholar; and Rosia aged 2. All of this third generation were born in Barnstaple.

William Moncrieff Jones

The man standing behind and to the left of Hannah is William Moncrieff Jones, an accountant, who is recorded in the 1881 Census as being 32, and living at 7 Alexandra Place, later Alexandra Road, with his wife Elizabeth, for whom no occupation is listed. They had no children. William and his wife were still at the same address in 1891, but by this time he was described as 'manager, bobbin net factory'.

When Miller Brothers became a limited company in 1892, William was named as secretary and manager[27], and the obituaries of the brothers indicate a close working relationship. He sat in the fourth carriage of mourners at the funeral of John May Miller in May 1899 with J. Hinde, mill manager, and W.H. Davie, and the floral tribute from himself and Elizabeth was signed 'with all love and respect'.[28] Alfred Miller had been taken ill with a cold while visiting his nephew in Nottingham and which developed into the bronchitis from which he died. As his condition had worsened on the Friday William had *left Barnstaple on Saturday morning, but when he reached Nottingham Mr Miller was already in a comatose condition.*[29] He was the only representative from Barnstaple at the cremation in Manchester and this, together with the other instances above, indicate his status both in professional and personal terms during a long association with the firm.

Next to William Jones is Bessy Harding. She worked as a lace mender, and appears in the 1881 Census as a fifteen year old living in Boden's Row. She is described as the daughter of Thomas Harding, the head of the household, but in the 1871 Census she is listed as Thomas's grand daughter aged 3. In 1881 Thomas and his wife had their own daughter Annie aged 19, a general domestic servant, and two other grandchildren, 9 year old Thomas Hill, and 5 year old Annie Giddy, both scholars, living with them.

Bessy Harding

Thomas Harding was taken on as a lace twister at the factory when John May Miller was recruiting replacements for the men involved in the lock-out dispute in 1874.[30] Prior to this he had been a mariner, and the marriage register of St.Mary Magdalene church records two instances of his daughters marrying seamen. In December 1870 one daughter Susan married George Mock, a mariner also of Boden's Row, and a fortnight later another daughter Ellen married Thomas Cook Hill, a mariner of Myrtle Place.[31] Perhaps Ellen and her husband were both at sea in 1881 when their son Thomas was staying with his grandparents. Susan's marriage was shortlived however as in January 1878 she appears again in the parish register as a widow on her marriage to James Fry, another mariner.

Bessy was still living in Boden's Row with her grandparents and working at the factory in 1891.

Bessy has her hand on the shoulder of Caroline Parsley. In 1881 she was 26 years of age and listed as a lace worker, living at home with her parents in Newington Street. Her father, John, and brother, Charles, were both carpenters and her other brother, Thomas, was a carter.

Ten years later she was still a lace mender and still in Newington Street with her

Caroline Parsley

99

parents. Thomas was now a horseman, and Charles had married and lived along the street with his wife and two children. Another brother, William, a shoemaker, had returned home to fill his place.

The second man in the bowler hat is labelled 'F. Hind' but this again, as with John Fear, is probably an error in transcription and he is almost certainly J Hinde, the mill manager with whom William

J Hinde

Jones shared the carriage at John May Miller's funeral. However there is no trace of him in Barnstaple in the 1881 Census.

Next to Caroline Parsley is the second Prust sister, and next again is a Mrs. Webster. The 1881 Census lists a Mary Webster aged 60, the wife of Henry Webster, a lace twister originally of Leicestershire. However the age of the woman in the photograph appears to be nearer that of

Mrs Webster

Henry's mother Ann, recorded in the 1861 Census as a 66 year old lace mender, living in Princes Street with her daughter Selina, who herself would have been around 88 by this time.

Behind them stands Ann Dyer, a lace mender aged 16 in 1881 and listed as a boarder, along with Charles Quick, a ten year old scholar, in the household of Elizabeth Westacott, a widowed lace mender of Newington Street. Elizabeth's children George, 20,

Ann Dyer

a lace twister, and Eva, 18, a lace mender, also show factory connections, whilst Jessie, aged 12 was a scholar. The 1891 Census lists Elizabeth A. Dyer, aged 25, a lace mender, as the unmarried head of a household in Reform Street. She had three young children living with her, William aged 4, Mary Ann aged 3 and Maud aged eight months, so she was probably working from home at this time.

Having now identified the occupations of most of those pictured in the photographs, the captions of 'Mill Staff' and 'Wareroom Staff' seem to be inappropriate. A more accurate pair of titles would be 'Blue Collar workers' and 'Women and White Collar men', as the clerical/managerial male staff are included with the women, and Bessie Dunsford, a lace twister, should have been included among the mill staff, certainly not in the wareroom.

Perhaps Bessie's true role might have offended Victorian niceties and whilst perfectly acceptable within the factory, might have been thought inappropriate to be recorded so publicly. Incidentally she is the only woman to wear a watch chain as the men do.

The Mill staff, so called, actually consist almost entirely of engineers and fitters, and only the Geens, who are described as 'machinists' and Philip Gardiner, a former lace twister now an overlooker, could possibly have been involved in the actual manufacturing process. Despite the accurate terminology, one wonders if the titles were transcribed from an original source or whether they are more modern additions.

Evidently there were many more employees than those pictured here, but whether more photographs were taken at the time or whether these workers, in many cases longstanding and trusted employees, were chosen to be representative of the workforce, we cannot know. However, whatever the circumstances of their origins, the old adage that every picture tells a story is certainly true many times over in their case.

Up until the 1960s bobbinnet was still being produced at the Barnstaple factory of John C. Small & Tidmas albeit not in the same quantities as previously. At that time nets were still being sent from the company's other production plants to Barnstaple for mending and it was not until the early 1970s that the final piece of net was despatched from the Barnstaple premises.

Up until this time the 'ware rooms' on the site of the present mill shop and offices were still inhabited by the menders on their low stools hand mending any holes in the nets so that when they were pinned out on the finishing frames they would not split. Every day long rolls of completed nets were manhandled across the yard, past an orchard, which was on the site of the present car park, and into the ware-house, while in the area above the present reception room the bobbins were being prepared on the cone-winding and cotton-waxing machines for loading on to the net making machines.

Again at this date the Works Manager in charge of net production would weekly deliver nets for mending to out-workers who could not collect their work from the factory. Mended nets would also be collected for onward despatch.

The increase in production and additional space requirements associated with the new warp knitting technology led to the end of net production in Barnstaple when the company decided to concentrate output at their other two sites at Perry Road in Nottingham and at Perry Street Works in South Chard.

David Dalziel, S&T. Barnstaple.

AFTERWORD

Although David Dalziel has sketched in recent changes at the Derby factory we have had to finish our main account in 1929 for a number of reasons. The absence of factory records has meant that we have had to rely on Census information for names of lace workers, and under the hundred year closure rule, the latest available is for 1891. The North Devon Journal is indexed up until 1895, so after this point articles involving Derby workers or residents may only be found by chance. The ending of the Miller family connection then seemed an appropriate place at which to stop. However, there is still much to be discovered and documented about the Derby area and the factory, but at present this exists only in the memories and photograph albums of those who lived or worked there. It would be wonderful to be able to follow up this volume with one which brought the history up to date, but this depends not only on the authors' time, but more particularly on people coming forward who are willing to share their stories and have them recorded for posterity. If you have enjoyed this book and are one of those people please get in touch with us via our publisher.

REFERENCES

Chapter 1

1. J. & B.Hammond *The Skilled Labourer 1760-1832* (London,1919) p.237
2. Patricia Wardle *Victorian Lace* (Herbert Jenkins,1968) p.222
3. Sheila Mason *Nottingham Lace 1760s to 1950s* (1994) p.273
4. Sherborne Mercury (SM) 25.2.1822 4c. Although all writers on Barnstaple give the date of opening as 1821 it is apparent from this report that the 1822 date is correct.
5. *Pigot & Co.'s London & Provincial Commercial Directory* (London,1823-4)
6. Mason (1994) p.273
7. North Devon Journal (NDJ) 5.11.1824 4a
8. NDJ 19.11.1824 4a
9. Rev.R.J.E.Boggis *A History of the parish and church of St.Mary Magdalene, Barnstaple* (Canterbury, 1915) p.167
10. H.W.Strong *Industries of North Devon* (Barnstaple, 1889) p.12. Reprinted with a new introduction by Barry Hughes (David & Charles, 1971)
11. SM 12.9.1825 4d
12. NDJ 24.8.1827
13. NDJ 15.11.1827 4c
14. NDJ 24.6.1830 4b-d
15. NDJ 17.10.1833 4b-d
16. NDJ 9.12.1830 4b-c; 16.12.1830 1c
17. NDJ 16.12.1830 4b
18. NDJ 23.12.1830 4b
19. NDJ 29.4.1830 4c
20. NDJ 2.1.1834 4b; 22.6.1837 4b
21. NDJ 11.6.1835 1e
22. NDJ 18.5.1837 4b; 1.6.1837 4b-c. The closure is also dealt with in the section on the Derby factory.
23. NDJ 19.4.1838 4c
24. NDJ 24.9.1840 2f
25. NDJ 16.6.1842 3a
26. NDJ 23.2.1843 3d
27. NDJ 27.6.1844 2f
28. NDJ 24.4.1845 2f
29. NDJ 12.7.1849 1b

30. NDJ 13.9.1849 5a. They were, however, used for something similar in 1855 when the Barnstaple Improvement Committee were undertaking the improvement of the Workhouse drainage and sewage pipes. Alderman Budd noted that the defective drains *had operated so prejudicially at the last visitation of cholera that the Guardians had been obliged to remove 50 of the children to Rawleigh factory for the recovery of their health.* – NDJ 28.6.1855 5d

31. NDJ 27.3.1862 1a

32. Strong (1889) p.13

33. Strong (1889) p.13; NDJ 26.4.1832 1a

34. NDJ 3.1.1828 4b

35. NDJ 9.2.1827 4a

36. NDJ 10.8.1827 1e

37. NDJ 1.11.1827 4a

38. NDJ 26.6.1828 4b

39. NDJ 25.8.1832 4b

40. NDJ 9.5.1833 4b

41. NDJ 19.9.1833 4b

42. NDJ 18.5.1837 4b

43. NDJ 14.12.1837 1b

44. NDJ 21.12.1837 4d

45. NDJ 4.1.1838 4c

46. NDJ 1.5.1845 3a; 15.5.1845 2f

47. The London Gazette 3.9.1841 as reprinted in *The London Gazette:Devon Extracts 1665-1850* edited by Marjorie Snetzler (Devon Family History Society, 1987)

Chapter Two

1. SM 31.10.1825 4d

2. SM 5.12.1825 4c

3. SM 12.12.1825 4c

4. Boggis (1915) p.168

5. NDJ 17.1.1828 4b

6. NDJ 22.5.1828 4c

7. Mason (1994) p.273

8. Boggis (1915) p.167

9. The London Gazette 14.11.1828

10. NDJ 29.12.1831 1d

11. NDJ 31.5.1832 4c

12. NDJ 21.6.1832 4a-d

13. NDJ 21.2.1833 4c; 19.7.1860 8a
14. David Thompson *Book of Remembrance or A Short History of the Baptist Churches in North Devon* (1885)
15. NDJ 20.6.1833 4b
16. NDJ 14.11.1833 4b
17. NDJ 21.11.1833 4c
18. NDJ 28.11.1833 4b-d
19. NDJ 28.11.1833 4d
20. NDJ 5.12.1833 4b-e
21. NDJ 19.12.1833 4b. This suggests files on the strike may well be held amongst the Home Office files in the Public Record Office.
22. NDJ 29.10.1835 4d
23. NDJ 7.1.1836 4b-c
24. NDJ 3.3.1836 4b; 31.3.1836 4b
25. NDJ 13.10.1836 1d
26. Strong (1889) pp.16 & 25
27. NDJ 10.11.1836 4b
28. NDJ 20.4.1837 1a-e
29. NDJ 27.4.37 1e
30. NDJ 18.5.1837 4b
31. NDJ 1.6.1837 1f
32. NDJ 1.6.1837 4b
33. NDJ 8.6.1837 4b
34. NDJ 8.6.1837 1f
35. NDJ 31.8.1837 4a
36. NDJ 7.6.1838 4d
37. NDJ 15.11.1838 2e
38. NDJ 6.2.1840 2e-f, 3a
39. NDJ 7.5.1840 3a
40. NDJ 4.11.1841 3b
41. NDJ 18.11.1841 3a
42. NDJ 2.12.1841 3b
43. NDJ 3.2.1842 2f
44. NDJ 10.2.1842 2e
45. NDJ 14.4.1842 3a; 5.5.1842 3c
46. NDJ 16.6.1842 3a
47. NDJ 27.10.1842 3e
48. NDJ 15.12.1842 3a
49. NDJ 2.3.1848 1b
50. NDJ 21.2.1850 4e

51. Original will held at Somerset House
52. The London Gazette 29.4.1845
53.   "   "   11.10.1846
54. NDJ 28.9.1848 2e
55. NDJ 5.10.1848 3d
56. NDJ 8.11.1849 5e
57. NDJ 29.11.1849 4d
58. NDJ 21.2.1850 4e
59. North Devon Record Office (NDRO) – 1096/8
60. NDJ 26.8.1852 4d
61. NDJ 30.9.1852 4c
62. NDJ 12.1.1854 5a
63. NDJ 23.2.1854 5a
64. NDJ 9.3.1854 5a
65. NDJ 31.1.1856 5b
66. NDJ 16.4.1857 5b
67. NDJ 5.11.1857 5a; 1.6.1899 5e
68. NDJ 5.1.1860 4e
69. NDJ 23.2.1860 5a
70. NDJ 1.6.1899 5e
71. NDJ 25.2.1864 5c
72. NDJ 12.4.1860 5b; 26.5.1860 5b
73. NDJ 28.6.1860 5e
74. NDJ 12.7.1860 5b

Chapter 3

1. NDJ 9.7.1863 5c
2. NDJ 21.11.1901 5d
3. NDJ 7.5.1903 5e-f
4. NDJ 9.7.1863 5c
5. NDJ 16.11.1863 6a-c
6. NDJ 15.2.1872 5a-b
7. NDJ 9.4.1868 5c; 1.6.1899 5e
8. NDJ 1.6.1899 5e
9. NDJ 24.8.1871 2a
10. NDJ 8.2.1872 5a
11. NDJ 15.2.1872 5a-b
12. NDJ 26.12.1872 4f
13. NDJ 18.8.1881 5c
14. NDJ 3.4.1873 5e
15. NDJ 26.6.1873 5b

16. NDJ 9.4.1874 5a
17. NDJ 10.9.1874 4f
18. NDJ 17.9.1874 5c
19. NDJ 24.9.1874 4f
20. NDJ 15.10.1874 4f
21. NDJ 29.10.1874 5a-c
22. NDJ 12.11.1874 5c
23. NDJ 12.11.1874 5c
24. NDJ 31.12.1874 5b
25. NDJ 28.1.1875 5c
26. NDJ 15.10.1874 4f
27. NDJ 1.7.1875 5f
28. e.g. NDJ 22.6.1876 5f, 12.7.1873 4f, 11.2.1886 5b
29. NDJ 8.4.1877 5d
30. J.M.Pedder *The Miller Family in the 19th.Century and the Derby Lace Factory* – manuscript in N.Devon Athenaeum p.8
31. Strong (1889)
32. NDJ 7.5.1903 5e
33. NDJ 7.5.1903 5e
34. NDJ 10.10.1901 8a-g
35. NDJ 7.5.1903 5f
36. NDRO – Barnstaple Town Council Box 4
37. J.M.Pedder p.16
38. Strong (1889) p.xv
39. NDJ 8.6.1899 3a
40. NDJ 21.11.1901 5d
41. NDJ 7.5.1903 5e & f

Chapter 4
1. NDJ 28.12.1906 2c
2. NDJ 30.12.1909 2d
3. NDJ 13.8.1914 2c
4. NDJ 13.8.1914 5c
5. NDJ 22.12.1914 7b & c
6. NDJ 20.12.1915 2a
7. NDJ 7.6.1917 5c
8. NDJ 4.11.1920 5g
9. NDJ 16.12.1920 3c-e
10. NDJ 30.12.1920 5c
11. NDJ 29.12.1921 3a

12. NDJ 30.6.1921 5b
13. NDJ 30.4.1925 5a
14. NDJ 14.5.1925 4g
15. NDJ 20.6.1929 5a & e
16. NDJ 25.7.1929 2f
17. NDJ 15.8.1929 5d
18. Letter from Nottinghamshire Museum of Costume & Textiles 27.6.1996
19. NDJ 3.10.1929 5b
20. NDJ 17.10.1929 5a
21. NDRO – Barnstaple Town Council Box 4

Chapter 5
1. Census returns from 1841 to 1891 are held in the North Devon Athenaeum
2. Strong (1889)
3. NDJ 18.11.1839 3a
4. NDJ 9.2.1843 3a
5. NDJ 13.4.1843 2e & f
6. NDJ 23.1.1834 4c
7. NDJ 30.1.1834 4b
8. NDJ 16.1.1834 4c
9. NDJ 7.3.1838 4d
10. NDJ 24.5.1838 4a
11. NDJ 11.9.1873 5b
12. Boggis (1915) pp.151,152 &174-5
13. Boggis (1915) p.154
14. NDJ 26.2.1829 4b
15. NDJ 7.5.1829 4b
16. NDJ 21.8.1833 4b
17. NDJ 22.9.1842 2f
18. NDJ 12.9.1833 4b
19. NDJ 3 10.1833 4b
20. NDJ 25.5.1827 4b; 5.2.1835 4d
21. NDJ 25.5.1827 4b
22. NDJ 5.2.1835 4d
23. NDJ 6.3.1834 4b
24. NDJ 6.3.1834 4b
25. NDJ 28.8.1834 4b
26. NDJ 13.8.1840 3b
27. NDJ 26.12.1833 4b-c

28. NDJ 4.11.1852 4e
29. NDJ 21.10.1852 5a
30. NDJ 13.1.1833 4b
31. NDJ 5.12.1833 1d

Chapter 6
1. NDRO – Barnstaple Parish Register
2. Boggis (1915) p.168
3. NDRO – B58 Z/1/1
4. NDRO – B59 Z/1/1
5. Census 1851 in North Devon Athenaeum
6. NDRO – B58 Z/1/2
7. NDRO – 320 IV/1/15
8. Census 1891 in North Devon Athenaeum
9. O.S.Devonshire Sheet XIII 1890
10. Boggis (1915) p.168
11. NDRO – 2654 A/64
12. NDRO – 774/2
13. NDRO – B253/5
14. NDRO – 774/1
15. NDRO – 702/6
16. NDRO – Barnstaple Castle Records 4867
17. NDJ 6.7.1843 3a
18. NDJ 21.6.1832 4a-d
19. NDRO – 1096
20. Boggis (1915) p.168
21. NDRO – 774/4
22. A.Dibben *Title Deeds* (The Historical Association,1971) p.8
23. W.F.Gardiner *Barnstaple 1837- 1897* (1897) p.131
24. NDRO – 705 c/7
25. NDRO – 705 c/15
26. NDRO – 705 c/17
27. NDRO – 705 c/25
28. NDRO – 705 c/27
29. NDRO – 705/c/18
30. *Billings 1857 Directory & Gazetteer of Devonshire*
31. NDRO – 702
32. NDRO – B58 Z/1/2
33. NDRO – B59 Z/1/1
34. NDRO – B58 Z/1/1. See ref.5 for Chapter 2 which seems to fit in with this observation.

35. NDRO – B58 Z/1/2
36. NDJ 10.5.1855 5c & d
37. NDJ 19.5.1853 5c; 2.6.1853 8a
38. NDJ 31.5.1855 5e
39. NDJ 8.6.1827 4b
40. Gardiner (1897) p.63

Chapter 7
1. The two photographs are reproduced in the illustrative
   section of this book. The originals still hang in the present
   day premises of S & T in Barnstaple.
2. NDJ 12.7.1883 4f
3. NDRO – St.Mary Magdalene Parish Register
4. NDJ 3.5.1917 8c
5. NDRO – St.Mary Magdalene Parish Register
6. NDJ 13.6.1833 4c
7. NDJ 27.10.1836 4b & c
8. NDRO – B58 Z/1/1 & 2
9. NDJ 15.2.1872 5 a & b
10. NDRO – B58 Z/1/2
11. NDRO – St.Mary Magdalene Parish Register
12. "   "
13. NDJ 26.9.1912 5d
14. NDJ 22.9.1842 2f
15. NDJ 1.12.1842 2f
16. NDJ 19.5.1853 5c; 2.6.1853 8a
17. NDJ 29.10.1874 5a-d
18. NDJ 20.7.1876 5f
19. NDRO – 1096/12
20. NDRO – St.Mary Magdalene Parish Register
21 "   "   "
22. NDJ 18.9.1924 8c
23. NDJ 29.4.1926 6d
24. NDJ 5.12.1912 8g
25. NDRO – St.Mary Magdalene Parish Register
26. NDRO – Barnstaple Parish Register
27. NDJ 7.5.1903 5e
28. NDJ 8.6.1899 3a
29. NDJ 7.5.1903 5e & f
30. NDJ 29.10.1874 5a
31. NDRO – St.Mary Magdalene Parish Register

INDEX

READER'S NOTES